GeoHibernica

GeoHibernica

The Irish Landscapes, Peoples and Cultures

Paul Lyle

LIVERPOOL UNIVERSITY PRESS

This book is dedicated to my family, friends and colleagues who have supported my literary endeavours over the past number of years. Their continued encouragement is greatly appreciated and without it I doubt whether I would have had the will to carry on. *Here's tae us! Wha's like us?*

First published 2025 by
Liverpool University Press
4 Cambridge Street
Liverpool
L69 7ZU

British Library Cataloguing-in-Publication data
A British Library CIP record is available

ISBN 978-1-78046-108-3 (hardback)
ISBN 978-1-78046-120-5 (epdf)
ISBN 978-1-78046-121-2 (epub)

Typeset by Carnegie Book Production, Lancaster
Printed in the Czech Republic via Akcent Media Limited

The manufacturer's authorised representative in the EU for product safety is:
Easy Access System Europe, Mustamäe tee 50, 10621 Tallinn, Estonia
https://easproject.com (gpsr.requests@easproject.com)

Foreword

The inspiration for this book came after a brief email correspondence with Mike Leeder in the period soon after the publication of *GeoBritannica: Geological Landscapes and the British Peoples* by Dunedin Academic Press (2017), which he co-authored with Joy Lawlor. In this exchange, Mike mentioned the possibility of writing a companion volume covering the landscape of Ireland. I confess it struck me as a good idea and one that I would not have minded undertaking myself. As things worked out, some years later, circumstances dictated that Mike was not in a position to carry out the project but generously allowed me to take on the concept of *GeoHibernica*, hence the current volume. I am happy to acknowledge the origin of the idea of looking at how the foundations of the landscape have been an influence on human activity and made the island of Ireland what it is today.

This book is a personal view of how the physical landscape of Ireland, the underlying geological foundation, has influenced human activities since the earliest settlers arrived on the island as the ice melted and receded.

Preface

It is probably true to say that interest in the Irish landscape has never been greater, not only among its residents, but further afield, as witnessed by the millions of visitors who come each year. There is a flourishing trade in books dealing with the archaeology, the geology and geography, the history, the literature of the island and there is apparently an insatiable appetite for television programmes for popular viewing that deal with the same topics. This curiosity is due to an increased awareness of the cultural significance of our surroundings to our physical and psychological well-being and an unease as to where a society, increasingly reliant on technology, is going. These misgivings are coupled with a growing concern about the future of the planet in the light of apparent rapid changes in climate leading to potentially catastrophic impacts on the environment. The last few decades have seen very profound changes worldwide in public attitudes to the natural environment and how it is perceived. In Ireland perhaps one of the most significant changes in public perception is the change in attitude to the large-scale exploitation of the country's peat bogs. Historically, these were regarded as an indigenous power supply, used to compensate for the lack of the sort of coal reserves found in Great Britain. Turf was an important source of domestic heating and cooking for thousands of years and since the early twentieth century was burnt on an industrial scale in the power stations of Bord na Móna. In a major reversal of Government policy since the beginning of the twenty-first century, these bog areas are now being conserved, re-flooded and recognized as a hugely important carbon and water storage facility.

GeoHibernica: The Irish Landscapes, Peoples and Cultures explores how Ireland's richly varied landscape – shaped by a diverse mix of rock types and geological ages – has influenced and sustained the ways of life of its inhabitants since the first settlers arrived during the Mesolithic era, around 10,000 years ago. As the ice sheets retreated some 13,000 years ago, the exposed bedrock and newly revealed land surfaces provided vital resources that enabled human settlement across the entire island. From the earliest working of the flint resources of the northeast and the first copper mines in the southwest the population expanded from the Neolithic era to the Bronze and Iron Ages and evolved from a hunter–gatherer society to one based largely on agriculture. The clearance of the original forest cover and the growth of the peat bogs invoked profound cultural changes, with influences seen even today. A modern interpretation of the geological history of Ireland provides a context for today's exploitation of metal ores and fossil hydrocarbons, including peat, in the light of concerns over atmospheric modification and the implications for climate change.

The book explores the overall cultural response to the landscape by society as a whole in the building of structures such as megaliths, high crosses and in the vernacular architecture, and also by individual artists, sculptors, writers and poets. The Irish landscape as an inspiration to painters and writers was highly significant during the early twentieth century when Ireland underwent major political change as it transitioned via Partition and the Irish Free State from being part of the United Kingdom to an independent country seeking a national identity. The socio-political impact of the surveying of the island by the Ordnance and Geological Surveys, which had strong military connections in their initial stages, is examined along with the influence of the chroniclers of the Irish landscape since Gerald of Wales in the twelfth century. The rise of naturalists such as Robert Lloyd Praeger at the end of the nineteenth century greatly encouraged later writers like Frank Mitchell and Estyn Evans to examine all aspects of the Irish landscape and its influence on the cultural environment of the island by looking at what Estyn Evans described as the 'habitat, heritage and history' of Ireland. This is a book for those who want to integrate the landscape they see around them today with that which confronted the earliest settlers arriving as the ice retreated – who also want to understand the changes that have occurred in the interim period and the cultural impact on society of the challenges as well as the resources that landscape provided.

Acknowledgements

I am pleased to acknowledge all those who have encouraged and supported me in the writing of this book. Mike Leeder gave me the concept of *GeoHibernica*, covering Ireland, following the publication of his excellent *GeoBritannica* book, which dealt with Great Britain, and which he co-authored with Joy Lawlor. They both read and commented on an earlier draft of the manuscript, and I am grateful for their perceptive and constructive comments. My profound thanks again go to John Arthurs who read and commented on the complete manuscript in its early stages and whose insightful criticism and comments greatly improved the content and structure of the book. I thank him also for allowing the use of the diagram illustrating the layers of the landscape, used in Chapter 1. Edward Ferguson gave me the benefit of his artist's insights and kindly allowed me to reproduce one of his paintings.

I am grateful to Dr Marie Cowan, Director of the Geological Survey of Northern Ireland (GSNI) for permission to use maps, diagrams and photographs from the GSNI archive and my thanks to Alex Donald for his considerable help in accessing this material. Dr Brian McConnell and Dr Siobhán Power of the Geological Survey of Ireland (GSI) are thanked for facilitating my use of maps and diagrams from the GSI archive. Gary Smylie played an invaluable role in IT matters pertaining to the book, for which I am deeply indebted. My thanks go to Anthony Kinahan of Dunedin Academic Press for his encouragement of the original concept. Dr Mike Simms of the Ulster Museum is thanked for permission to use a photograph of the fossil tetrapod trackway on Valencia Island, Co. Kerry. My thanks to Matthew Stout and the editors of the *Atlas of the Irish Rural Landscape* for permission to publish maps and diagrams. Liam McCaughey of the Belfast Naturalists' Field Club kindly supplied the image of their 1868 field trip to the Giant's Causeway, as well as supplementary information on the identity of some of the participants. My thanks to Donna Rowley for permission to use her digital image of the Giant's Causeway used in Chapter 6.

My thanks to those other artists, Emma Cownie, Doreen Drennan, Colleen Frew and Patricia Curran Mulligan, who graciously gave me permission to reproduce their work.

My family in Northern Ireland and Scotland have as usual played a key role in supporting and encouraging my writing and this is greatly appreciated. Sylvia Lyle, as always, has exerted considerable influence on the content of the book while also undertaking much of the early proofreading. Her constant encouragement and support are invaluable, and I am forever grateful.

I would also like to thank Rachel Chamberlain and all at Carnegie Book Production, and the team at Liverpool University Press, Michael Ainsley, Rebecca Graham and Patrick Brereton, for their editorial and design skills. These contributed significantly to the final version of the book.

Notwithstanding the support, encouragement and assistance detailed above, any errors or omissions are my responsibility.

Contents

Illustrations

Who can see the green earth any more
As she was by the sources of Time?
Who imagines her fields as they lay
In the sunshine, unworn by the plough?
Who thinks as they thought,
The tribes who then roam'd on her breast,
Her vigorous, primitive sons?

The Future, Matthew Arnold (1822–88)

1

Introduction

Figure 1.1 is a painting by Belfast-born artist Paul Henry. It depicts a scene in Connemara dominated by a mountainous background with cottages in the foreground. Henry was famous by this stage in his life for paintings of the countryside and people of the west of Ireland. They often, as here, depicted impressive mountains backed by towering cumulus clouds with homely features such as thatched cottages and turf stacks in the foreground. Such scenes were used as a basis for posters in the 1920s and 1930s by British and Irish railway companies as a means of promoting tourism to areas in Ireland such as the Wicklow Mountains, Killarney, Donegal or Connemara. They were intended for both the UK and Ireland markets and also to attract visitors from abroad. The effect of such images was to establish a standard scenic view of Ireland that extended beyond tourist literature to government and other publications and was to endure

Figure 1.1 *A Connemara Village*, by Paul Henry. Courtesy of the National Gallery of Ireland.

long after the importance of the railway companies had declined in the face of competition from the motor car. The impression of Ireland portrayed by these posters was clearly limited and specific to certain parts of the country, those regions which were considered to be more attractive scenically. This leads us to consider what the term 'landscape' as used in the title of this book means in the context of the whole island of Ireland.

It is possible to find any number of definitions of 'landscape'. 'All the visible features of an area of land, often considered in terms of their aesthetic appeal' (*Oxford English Dictionary*). 'A portion of territory that can be viewed at one time from one place' (*Merriam-Webster Dictionary*). 'The area and features of land that can be seen in a broad view, especially when they form a particular type of scenery' (*Chambers Dictionary*).

The title of the book refers to the Irish landscape, along with its people and culture. The type of image portrayed by Paul Henry and used as advertising by the railway companies is often described as 'landscape', but since, for example, the west of Ireland is not representative of the whole of the island, there is a need to consider exactly what is meant by the term. Human interaction with the environment has produced a range of distinctive 'landscapes', which reflect the diverse *physical* properties of the island and the complexity of the various *cultural* influences.

In any discussion of the concept of landscape it is helpful to consider the origin of the term. Most definitions refer to the visible features of an area of land and its landforms and how they integrate with natural or man-made features. Landforms can be defined as features on the surface of the Earth, such as plains, plateaux and mountains, or smaller-scale features, such as hills, valleys or river terraces. Many definitions refer specifically to the aesthetic appeal of the area. The modern form of the word with this connotation of scenery or aesthetic appeal appeared in the late sixteenth century when the word *landschap* was introduced by Dutch painters, who used it to refer to the natural features of a landscape in terms of their appearance – especially if regarded as spectacular or picturesque. Additional terms, such as seascape, city-scape or town-scape are recent variations on landscape.

In 1975, the English geographer Jay Appleton proposed in his book, *The Experience of Landscape*, reasons why certain landscapes are perceived as beautiful and others as bleak. Evolving from our hunter–gatherer ancestors we came to recognize symbolic aspects of the landscape – forests and other vegetation, cliffs and chasms, vantage points and open grasslands. He concluded that most people have an inborn desire for environments that allow the capacity to observe without being seen – in other words the ability to assess potential threats from a place of safety. On this basis, he proposed the prospect–refuge theory of human aesthetics, that taste in art is 'an acquired preference for particular methods of satisfying inborn desires'. These desires are for environments that allow the possibility of observation without being seen while being able to assess threats from a place of safety – hence the prospect (observation)–refuge (safety) theory. With this device, Appleton hoped to bridge the gap between the scientific examination of the landscape (which is often driven by practical considerations such as the search for resources) and the artistic interpretations more concerned with aesthetic aspects.

It is possible to separate the *physical* landscape from the *cultural* landscape. The former is the realm of the geomorphologist and the science of geomorphology. This seeks to understand why landscapes look the way they do and to understand the physical and chemical processes that operate at or near the Earth's surface. While these surface processes are paramount in sculpting the surface landforms, there are deeper geologic processes that cause tectonic or crustal uplift and subsidence leading to earthquakes, volcanic eruptions, mountain-building, deep sedimentary basins and sea level changes. Overall, the results of these processes are acted on by climatic factors.

In contrast, a cultural landscape, as defined by the World Heritage Committee of UNESCO (The United Nations Educational, Scientific and Cultural Organization), is defined as 'the cultural properties that represent the combined works of nature and man'. These can be:

1. A landscape designed and created intentionally by man.
2. An 'organically evolved landscape', which may be a fossil or relict landscape or a 'continuing landscape'.

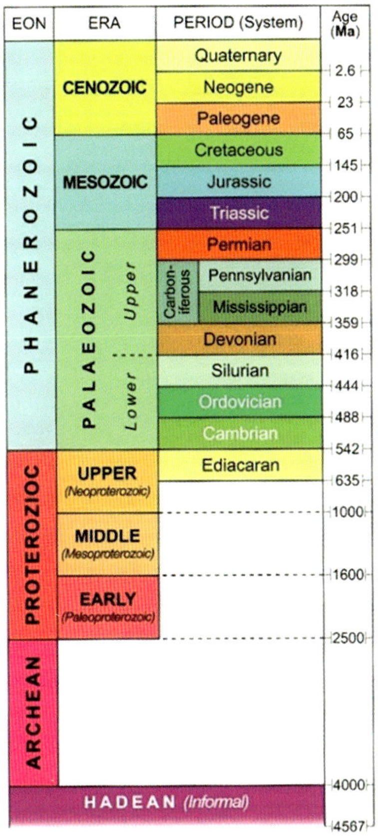

Figure 1.2 The main subdivisions of the stratigraphic column with approximate ages given as Ma (millions of years).

3. An 'associative cultural landscape', which may be valued because of the 'religious, artistic or cultural associations of the natural element'.

This book aims to examine the response by society to the environment in Ireland as the ice melted and retreated and the landscape evolved away from the post-glacial phase. As the population of Ireland steadily increased came the development and spread of agriculture, and eventual industrialization, and the environment we are familiar with today. The term 'landscape' as used in this book is deliberately interpreted in a very broad way. Earlier studies, such as E. Estyn Evans's *The Personality of Ireland* (1973), included oral traditions, beliefs, languages and artistic endeavours, and F. H. A. Aalen, Kevin Whelan and Matthew Stout's *Atlas of Irish Rural Landscape* (1997), explored those features that give the Irish countryside its character, features that included cultural elements, such as settlement patterns, archaeological monuments, communications, mines and factories. In this book I have endeavoured to include as wide a range of influences and factors as possible within the scope of a single volume by a single author.

The people who live in Ireland, or who visit it for any reason, can enjoy some of the most diverse scenery anywhere in the world, with spectacular coastlines, mountains, loughs, rivers and fertile lowlands. The variety of landscape reflects a number of influences, but primarily it is due to the remarkable range of age and composition of its bedrocks. On a relatively small island on the western fringes of Europe there are rocks from every division of the geological column. From the oldest on the island of Inishtrahull off the north coast of Donegal (*Proterozoic*, around 1.8 billion or 1.8 Ma old) to sediments deposited since the retreat of the glaciers of the last Ice Age (*Quaternary*, around 10000 or 10 ka old),[1] there are rock types ranging from highly contorted metamorphics, through igneous rocks such as granites (*Devonian* and *Paleogene*) and basalts (*Palaeogene*), to sedimentary rocks such as the sandstones and mudstones of Co. Down (*Ordovician* and *Silurian*), to the extensive limestones of the Burren area in Co. Clare (*Carboniferous*). Figure 1.2 shows the main subdivisions of the stratigraphic column, with approximate ages given as millions of years (Ma).

Matthew Arnold's poem *The Future*, quoted at the head of the chapter, was written in the middle of the nineteenth century, when there was an increasing awareness and knowledge of environmental matters. Although written by an English poet and presumably in the context of an English landscape, it is equally relevant to Ireland. The theme is of a former pristine state of the Earth being changed by modern life. It harks back to a former state when the earliest settlers were making their mark on an emerging landscape. In the subtitle of his classic study of Irish culture, *The Personality of Ireland*, mentioned earlier, Estyn Evans put forward the concept of 'habitat, heritage and history'. He then went on to produce what he described as an 'anthropogeographic view' of Irish history and culture that still ranks as one of the most comprehensive fusions of geology, geography, archaeology, anthropology and history ever written about the island.

Estyn Evans's view of habitat embraced the total environment; he defined heritage as the unwritten

1 Ma = 1 million years; ka = 1 thousand years.

Figure 1.3 The transformation of the natural landscape.

Figure 1.4 Layers of the landscape. Courtesy of J W Arthurs.

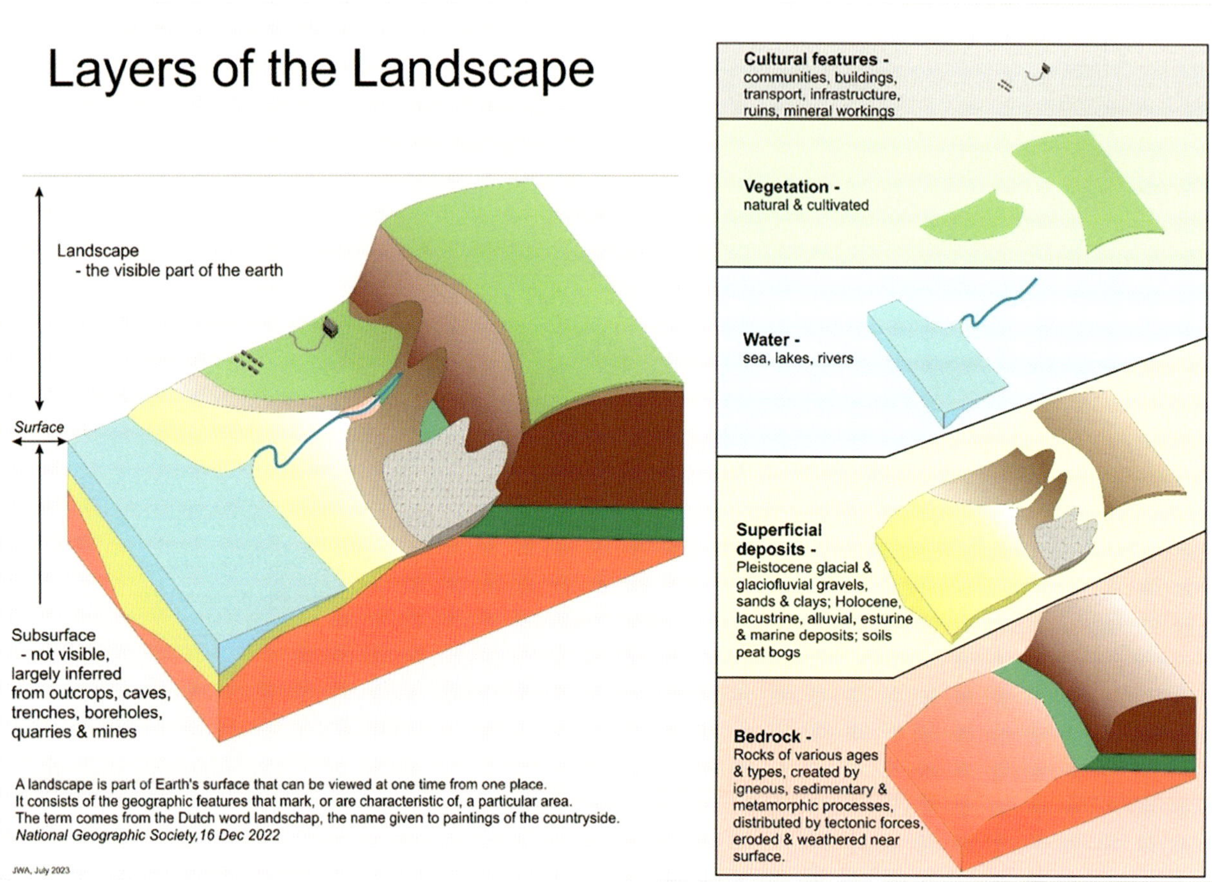

segment of human history comprising the physical, mental, social and cultural inheritance from the days of the earliest settlers onwards, and he considered history as the written record of past events. Heritage included oral traditions, beliefs, languages and artistic endeavours. Aalen, Whelan and Stout's *Atlas of the Irish Rural Landscape* sought to explore the physical and human factors that give the landscape its distinctive character, with an emphasis on the human elements, such as settlements, archaeological and historical monuments, woodlands, bogs, fields and buildings. The landscape we see today is the result of the bedrock geology of an area and its interaction with the forces of erosion and deposition by wind, water and ice that have carved and modified them throughout geological time. This can be considered to be the *natural environment*. However, since the retreat of the ice at the end of the last glacial event some 13,000 years ago, these natural

Figure 1.5 Poulnabrone dolmen or portal tomb, Co. Clare. Courtesy of Jon Sullivan. https://commons.wikimedia.org/wiki/File:Paulnabrone.jpg.

processes have been enhanced by additional changes brought about by successive generations of people who have settled the island since their first arrival, probably around 8,000 years BP.[2] This has produced a *cultural environment*, modified by advances in technology and changes in social organization (Figure 1.3). The landscape we see today is therefore many-layered, with some of those layers remaining from the earliest times, while others result from recent human activities (Figure 1.4).

The bedrock represents the fundamental layer on which all the other layers are laid. In Ireland, the bedrock is diverse both in terms of the rock types it consists of and also in the ages of the rocks. Superficial deposits are laid on top of the bedrock and comprise the glacially derived sands, gravels and clays, and the subsequent river, lake and estuarine deposits and peat bogs. The bedrock and superficial deposits are subject to modification by the *hydrosphere* – the seas, lakes and rivers of the Earth – and water is essential for the growth of vegetation, both natural and cultivated. Superimposed on all of these layers are the cultural features resulting from human activities. These include settlements, transport and other infrastructures, mining and quarrying impacts and cultural monuments.

Figure 1.5 shows the unusually large Poulnabrone *dolmen*, or *portal tomb*, in the Burren region of Co. Clare. The dolmen consists of a number of upright slabs known as *portal stones* or *orthostats*, on which rest a near horizontal *capstone*. The structure is dated to the *Neolithic* period, with radiocarbon dating of the bones of individuals buried there giving an age of 4200 BCE to 2900 BCE. It is located on an extensive area of bare rock within an area of *karst* landscape. Karst is a term used across the globe to describe the distinctive landforms that develop on rocks such as limestone, which are readily dissolved by rain and groundwater. One of the characteristics is the existence of areas of bare rock, referred to as *limestone pavement*. Such an area can be seen in the background of Figure 1.5.

Excavations at Poulnabrone in the 1980s uncovered the remains of 33 individuals, and radiocarbon dating of their bones indicates that the tomb was in continuous use between 5,200 and 5,800 years ago. The bones show signs of wear, indicating a life of hard physical labour, and one hip bone has the tip of an arrowhead embedded in it, suggesting conflict existed during that time. Evidence of creativity and craftsmanship was found in the form of a decorated neck pendant, beads and pottery fragments. The remains were in a disarticulated state suggesting the bodies had been treated elsewhere and then moved to the dolmen for final burial.

Poulnabrone is one of more than 1,500 such monuments, known as *megaliths*, found in the Irish landscape. They are a component of the major changes that were being made on the landscape during the Neolithic as the climate recovered from the Ice Age and the forests were being cleared as agriculture expanded. These monuments were erected for a variety of reasons – including reminders of past events, as territory markers and as part of the religious and ceremonial beliefs of the society that constructed them. Monument building enhances the significance of locations, which leads to a change in the perception of that landscape. They represent a desire to make an impression on their surroundings that indicate an integration of the physical and cultural environments, both of which were evolving. It seems that Neolithic society was beginning to see itself as part of a time continuum that included not just themselves and their ancestors, but future generations as well. The monument is constructed of slabs of limestone that were likely to have been sourced

2 BP = (years) before present.

locally, thus it could be said to be *in* the landscape but also *of* the landscape.

In his *Landscapes of Neolithic Ireland* (2000), Gabriel Cooney pointed out the difference between monuments that are built to blend in with the existing landscape features and those that are designed actively to create new and different spatial patterns. Poulnabrone, with its high visibility, clearly falls into this latter category. The building of such structures was amongst the first collective responses, leading to other societal activities, as well as efforts by individual artists, sculptors, writers and poets.

There is clear evidence of human settlement in Ireland before the Neolithic Period. In the preceding Mesolithic Period, around 9000 years ago, some of the earliest inhabitants of the island lived at Mount Sandel near Coleraine in Co. Londonderry. These people lived in circular huts made from saplings and enjoyed an abundant and varied food supply, including salmon and wild pig, supplemented by nuts and berries in the autumn. A supply of the mineral flint was an important factor in their existence and was used to manufacture a range of tools, including axes, scrapers and needles. Recent evidence has suggested that settlement may have begun earlier than the Mesolithic, in the preceding Palaeolithic Period, possibly as far back as 12,500 years ago. These early inhabitants made little impression on the landscape in the several thousands of years of their existence. They were hunter–gatherers, who lived on the coast and in clearings in the forest. This minimal impact is in stark contrast to the Neolithic, where the building of megalithic monuments was one radical change in the landscape, while the other was the clearance of the forests and the development of agriculture. Monument-building requires a complex social structure capable of organizing communal effort. This may have happened because the development of a more settled way of life based on agricultural practices allowed more leisure time than that available to the earlier hunter–gatherers of the Mesolithic.

Figure 1.6A Satellite image of the area around Strangford Lough, Co. Down. Contains modified Copernicus Sentinel data 2018, CC BY-SA 3.0 IGO https://creativecommons.org/licenses/by-sa/3.0/igo/deed.en, via Wikimedia Commons.

Figure 1.6C Dolerite sill in Trias sandstone.

Figure 1.6B Scrabo Hill and Tower from the south.

Figure 1.7 Mount Stewart House, Co. Down, a nineteenth-century demesne, seat of the Marquis of Londonderry.

Alternatively, agriculture and monument-building may have proceeded as parallel processes. Resolution of this conundrum is still a matter of some debate in archaeological circles, but what is clear is that the Neolithic marks a radical change in what had been the norm previously, and the view of the environment by the island's population was changing. This was the beginning of a process of landscape evolution that is continuing to this day whereby the landscape could be modified and changed for the benefit (or not) of society.

Case Study: Strangford Lough and Newtownards, Co. Down

The composite and complex nature of the modern landscape is very well illustrated by the region around Newtownards and the northern end of Strangford Lough, on the east coast of Co. Down, in northeast Ireland (Figures 1.6A–C). Scrabo Hill is a prominent topographical feature topped by Scrabo Tower, a Gothic-style memorial to the Third Marquis of Londonderry, erected in the 1850s, and clearly visible over a wide area. Scrabo is composed of reddish sandstones

Figure 1.8 Remains of the buildings at Nendrum, Co. Down.

of Triassic age that have been preserved from erosion by a capping of a hard igneous rock, *dolerite*, which was intruded to form a *sill* during the same volcanic episode that formed the Antrim basalts further north. The hill forms a *crag and tail*, Figure 1.6 B, a glacial feature fashioned when moving ice meets a resistant mass of rock (the crag), in this case the hard dolerite intrusion, and flows around it. This leaves a gently sloping deposit of glacial sediment (the tail) on the lee side. The ice here was flowing from right to left. While the crag and tail feature has traditionally been explained in this way, recent research on Scrabo suggests that the direction of ice flow may, in fact, have been in reverse.

The Triassic sandstone has been quarried since Anglo-Norman times, when, amongst other things, the sandstone flags were used as coffin lids. Since then, the rock has been employed as a building stone, for both prestige buildings in Belfast and Dublin and in the vernacular architecture of nearby Newtownards and district. The landscape around Newtownards and Strangford Lough is complex and shows multiple influences. The name *Strangford* is Old Norse, meaning *strong sea inlet*, and is an indication of Viking activity in the area. That Scrabo Hill is a special site is seen from the positioning of the memorial tower and the presence of Iron Age huts near the summit. In recent years, the hill has been used for open-air worship by the local Christian churches on Easter Sunday. Current uses of the land around the hill include a golf course, Newtownards airport, a modern road network and some of the best farmland in Ireland. The disused quarries now serve as an important nature reserve and country park. Within a few miles of the hill are the Marine Nature Reserve of Strangford Lough; Mount Stewart House and estate (a prime example of a nineteenth-century demesne Figure 1.7); former monasteries at Greyabbey and Nendrum; and Newtownards airport.

Nendrum Abbey and Greyabbey, the ruins of which lie on opposite sides of the head of Strangford Lough, show a contrast in monastic settlement types and their use of the landscape. Nendrum developed around the seventh century and consists of three enclosures. The inner contained the church and the graveyard, the middle enclosure had round houses, with living areas for the monks, while the outer precincts had craft workshops, such as blacksmith forges (Figure 1.8).

Figure 1.9 Elaborate carving on west door, Greyabbey, Co. Down.

Nendrum Abbey also operated fish traps in the nearby lough and constructed an elaborate tide mill using local sandstone and Mourne granite millstones. What would have been quite a bustling settlement, engaged fully with the local population, contrasted with Greyabbey a few miles away, built by the Cistercians, a contemplative order (Figure 1.9). It was founded in 1193, by John de Courcy's wife, Affreca (daughter of Godred Olafsson, King of the Isles), as a daughter house of Holmcultram Abbey in Cumbria. It had declined by the late Middle Ages and was dissolved in 1541. Cistercian monasteries were generally grouped around a quadrangle, and Greyabbey boasts an elaborately carved west door.

Many of the changes in the Irish landscape, both naturally and culturally derived, can be seen here in microcosm. The physical characteristics of the landscape, the bedrock geology and the processes of weathering, erosion and deposition provide the framework on which the cultural, social and industrial development of the region is played out. Recognition of the various distinctive physical landscapes that make up Ireland provides a context for the raw materials used by industry and in the construction of the historic and vernacular buildings, and how architects, artists, sculptors and writers have responded to these influences.

2

The Foundation of the Irish Landscape: The Natural Regions

Ireland can be subdivided into a number of *natural* or *physical* regions. A satellite image shows a central plain or lowland area surrounded by coastal hill and mountain ranges, with the western coast characterized by prominent bays marked by headlands and peninsulas (Figure 2.1). The interior contains several major rivers and lakes (*loughs*); the largest river is the Shannon, which discharges into the Atlantic at the Shannon Estuary, via Loughs Allen, Ree and Derg. The largest lough is Lough Neagh, in the northeast, which flows

Figure 2.1 Colour satellite image of Ireland. Jeff Schmaltz, Public domain, via Wikimedia Commons.

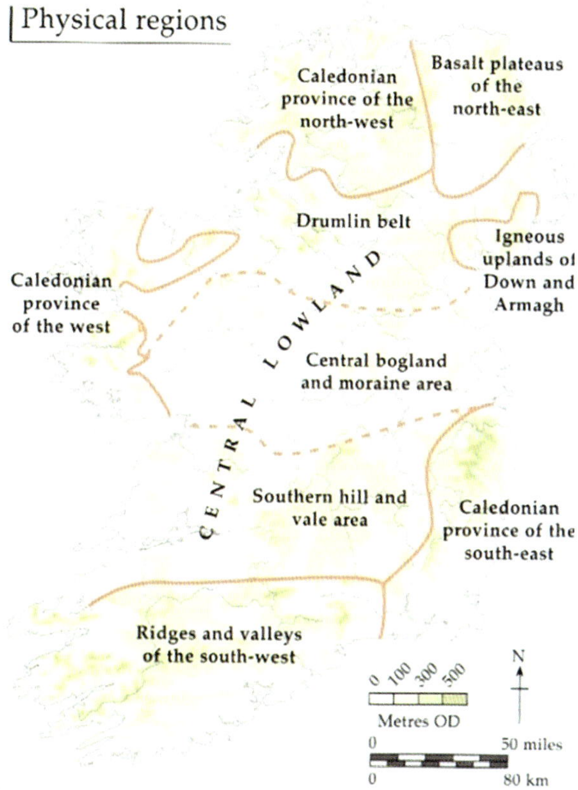

Figure 2.2 Physical regions of Ireland. Courtesy of *Atlas of the Irish Rural Landscape.*

into the Atlantic on the north coast via the river Bann. The approximate boundaries of these regions are shown in Figure 2.2.

The Physical Regions of Ireland

1. The Caledonian highlands
2. The northeastern Palaeogene basalts
3. The drumlin belt

4. The eastern granitic uplands, Down and Armagh
5. The Central plain with glacial sands, eskers and bogland
6. The southeastern Caledonian province and Leinster granite batholith
7. The Munster ridge and valley province

The Bedrock Geology: Summary

The subdivision of the island into these natural/physical regions is attributable to the underlying bedrock and the erosional and depositional effects of the most recent phase of the Ice Age showing the bedrock geology of Ireland (Figure 2.3). The boundaries shown in black lines approximate to those of Figure 2.2. Shown in Figure 2.4 is the key to the colours of the main rock types along with the main environmental changes and tectonic events.

In the west and northwest, from Co. Galway to Co. Donegal, the rocks are predominantly high-grade metamorphic rocks of Precambrian age (see the stratigraphical column), with affinities to rocks of similar types and age in the Scottish Highlands. The Precambrian metamorphic rocks found in Donegal continue eastwards under the basalt plateau and crop out on the northeast corner of the island. The northeast corner comprises the Palaeogene (formerly Tertiary) Antrim Plateau lavas (see stratigraphical column), with connections to similar lava fields in western Scotland, notably the islands of Mull and Skye. Immediately underlying the basalts are thin successions of Mesozoic rocks, most prominently Triassic sandstones and Cretaceous chalk. South of the lava plateau is a belt of Ordovician and Silurian rocks, mostly sandstones and mudstones, which extends southwest to Co. Longford and Navan in Co. Meath, with affinities to the Southern Upland Province of Scotland. The Mourne Mountains and the Slieve Gullion/Carlingford areas are Palaeogene intrusive complexes and along with the Newry and Slieve Croob granites of Devonian age form a distinct region of higher ground in contrast to the more fertile lowlands surrounding them (see stratigraphical column and bedrock map, Figures 2.3 and 2.4).

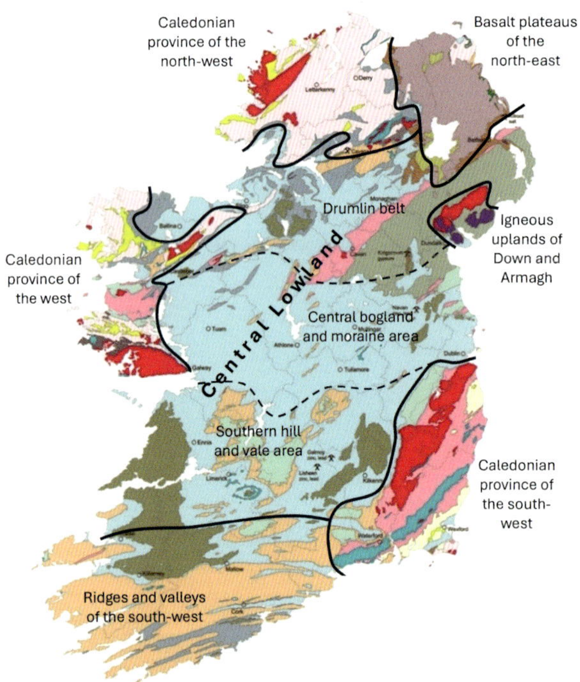

Figure 2.3 Bedrock geology of Ireland. GSI.

Over the Central Plain is a comprehensive coverage of Carboniferous limestone, forming a comparatively fertile region dominating the middle of the island. The Bog of Allen is a large, raised bog in this central area, between the rivers Liffey and Shannon. On its western edge, the limestone is uncovered and forms the distinctive karst landscape as seen in the Burren region of Co. Clare. Extending from the Central Plain is the Southern Hill and Vale Province, Carboniferous shales that have been eroded in places to form flat-topped mountains, and in places, deeper layers of slate and shale have been folded to emerge as mountains, such as the Slieve Bloom and Galtee ranges. Along the east coast, and south of Dublin as far as Counties Waterford and Wexford, are Ordovician and Silurian sediments. In contrast to their northern counterparts, which, as noted, have Scottish affinities, these are similar to rocks in Wales. The Wicklow Mountains are granites of Caledonian age intruded into the Ordovician and Silurian sediments. In the southwest of the country, in Co. Kerry, is an area of deformed Devonian rocks forming mountain ranges such as the MacGillycuddy's Reeks (Figure 2.4). For an explanation of rock terms, see the Glossary.

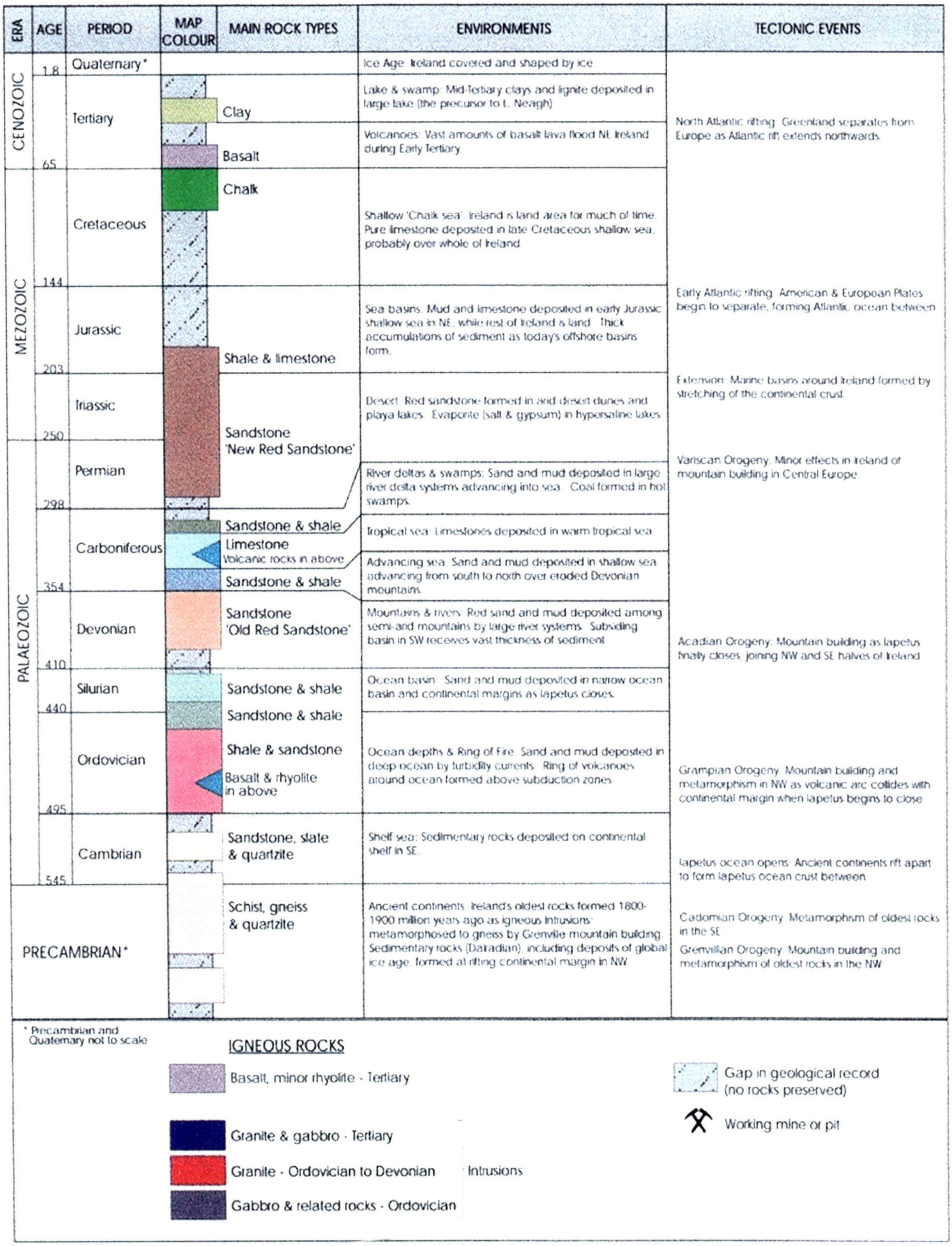

Figure 2.4 Annotated stratigraphical column showing main environmental changes and tectonic events.

Characteristics of the Physical Regions of Ireland

The Caledonian Highlands or North-Western Caledonian Province

In this section the satellite images and geological maps are taken from Figures 2.1 and 2.3 respectively.

This region comprises the Palaeozoic rocks of counties Tyrone, Londonderry and Donegal in the northwest corner of the island and is continued further to the south and west to cover the counties Mayo and Galway (Figures 2.5 A–C). The rocks are characteristically hard quartzite metamorphic rocks with affinities to rocks of similar age in the Scottish Highlands. They were formed in a mountain-building episode or *orogeny*, generally referred to as the *Caledonian orogeny*. An orogeny is the process in which a section of the Earth's crust is folded and deformed by compression to form a mountain range. Separation between the two parts of the region is due to the softer, more easily eroded Carboniferous rocks around Donegal Bay. These Caledonian rocks continue eastwards, at depth, under the Antrim Plateau, and occur again at the surface on the northeast corner of the island.

Figure 2.5A Satellite image, Caledonian Highlands.

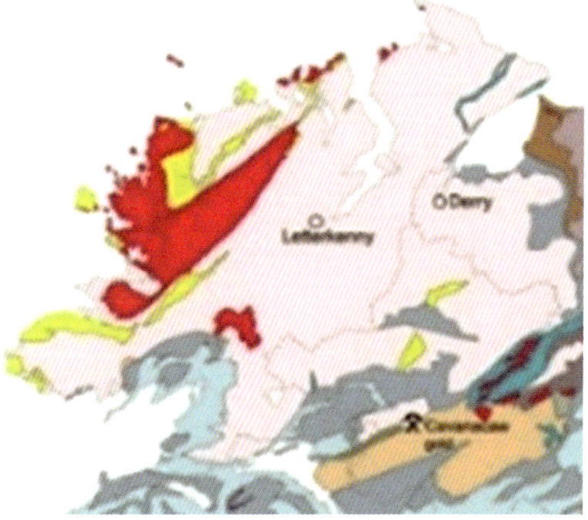

Figure 2.5B Bedrock geology, Caledonian Highlands.

Figure 2.5C Dalradian metamorphic rocks, Slieve League cliffs, Co. Donegal.

The North-Eastern Palaeogene Basalt Province

The basalt province is part of the wider North Atlantic Igneous Province, formed by the extension northwards

Figure 2.6A Satellite image of the northeastern Palaeogene basalt province.

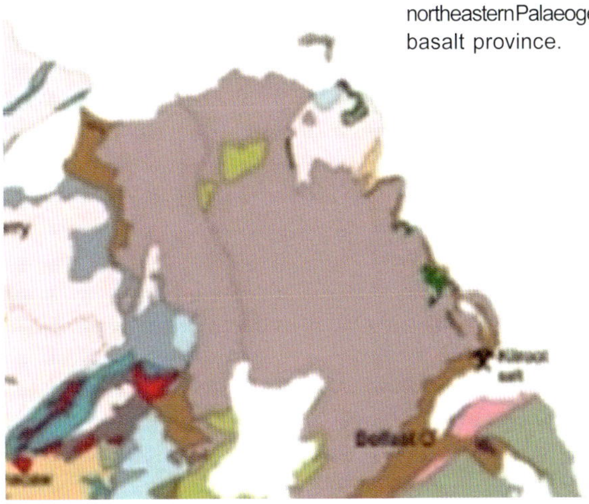

Figure 2.6B Bedrock geology of the northeasternPalaeogene basalt province.

Figure 2.6C Slemish Mountain, Co. Antrim, feeder plug to the Antrim lavas.

of the Atlantic Ocean as Greenland separated from Europe in the Palaeogene, formerly known as the Tertiary Period (Figures 2.6A–C). The hard, resistant lavas form spectacular cliffs, and the plateau is dissected by a number of valleys or glens, and flows dip generally into the basin of Lough Neagh to the west and south. Immediately underlying the basalts are thin successions of *Mesozoic* rocks, most prominently Triassic sandstones, indicative of a desert environment, and Cretaceous chalk formed in a warm shallow sea that covered much of Europe.

The Drumlin Belt

Drumlins are small oval-shaped hills of *glacial drift* (the general term for glacially derived sediments) that formed under a glacier or ice cap and aligned in the direction of ice flow. They occur in many places in Ireland but are particularly prominent in a belt from Co. Down on the east coast through to the west coast at Clew Bay in Co. Mayo (Figures 2.7A–C). The term 'basket of eggs topography' is sometimes used to describe the distribution of swarms of these hills whose shape is like an upturned boat, or half an egg. The lower ground

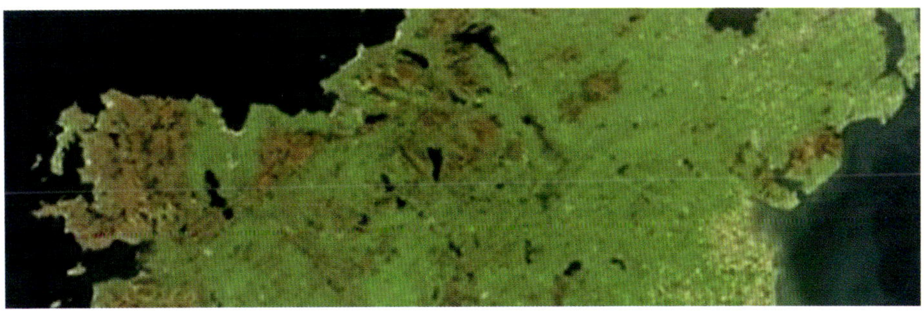

Figure 2.7A Satellite image, the drumlin belt.

Figure 2.7B Bedrock geology, the drumlin belt.

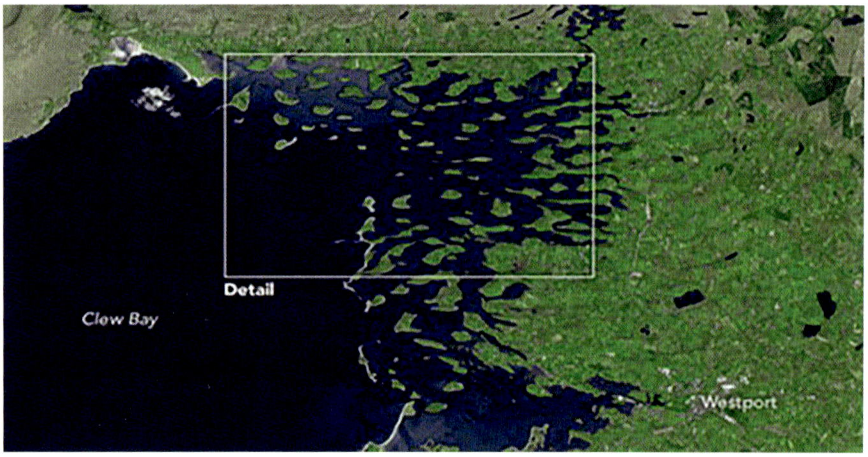

Figure 2.7C Drowned drumlins, Clew Bay, Co. Mayo.

between the hills tends to become waterlogged and the result is a tendency to small farms, agriculturally poor and often struggling for viability. Strangford Lough in Co. Down in the east and Clew Bay in Co. Mayo in the west are two locations where drumlin areas have been flooded by the sea, and the hills now occur as small, rounded islands. The drumlin zone occurs principally over a belt of Ordovician and Silurian rocks, mostly sandstones and mudstones, which extends from Co. Down in the east, through Co. Armagh, Co. Monaghan and Co. Longford. These rocks have close affinities to the Southern Uplands area in Scotland.

Figure 2.8A Satellite image, eastern granitic uplands.

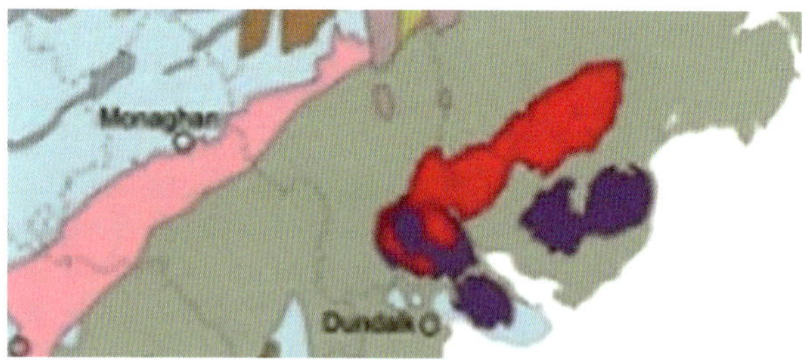

Figure 2.8B Bedrock geology, eastern granitic uplands.

Eastern Granitic Uplands

This upland region covering parts of Counties Down, Armagh and Louth consists of the mainly granitic rocks of the Mourne Mountains and the Newry granodiorite intrusion of Slieve Croob in Co. Down, the gabbroic intrusion of Slieve Foy on the Cooley Peninsula beside Carlingford Lough in Co. Louth and the basaltic/granitic intrusion of Slieve Gullion in Co. Armagh (Figures 2.8A–C). The Mourne, Carlingford and Slieve Gullion complexes are all Palaeogene in age, a late phase of the activity responsible for the extrusion of the basalts further to the north. The Newry complex is older, part of intrusive activity associated with the Caledonian orogeny in Devonian times.

Figure 2.8C Slieve Foy, Co. Louth, part of the Palaeogene Carlingford igneous complex. https://commons.wikimedia.org/wiki/File:Slieve_Foy._Carlingford._Ireland_-_panoramio.jpg, CC BY-SA 3.0 https://creativecommons.org/licenses/by-sa/3.0, via Wikimedia Commons.

Figure 2.9A Satellite image, central plain, with glacial sands.

The Central Plain with Glacial Sands and Eskers

The Central Plain is underlain mainly by limestones of Carboniferous age, which are covered by raised bogs such as the Bog of Allen, glacial sands and gravels and eskers (Figures 2.9A–C). Eskers are long winding ridges of sands and gravels left by retreating glaciers. On the western edge of this Carboniferous limestone outcrop, the glacial sediment cover is sufficiently thin to expose the original limestone land surface, which is now characterized by classic karst features. Karst is a topography formed by the dissolution of water-soluble rocks such as limestone and typically has underground drainage systems with sinkholes and caves.

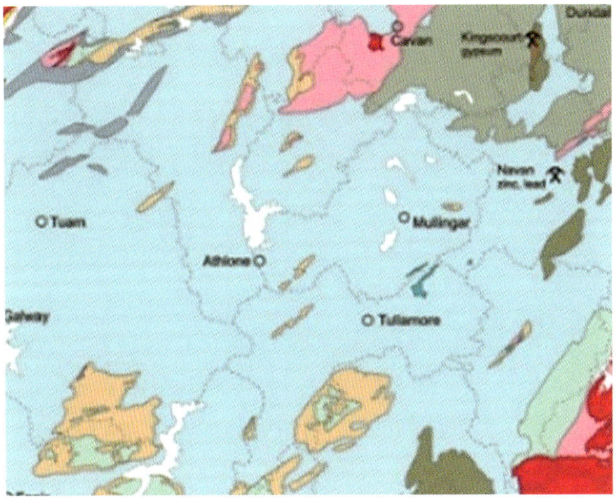

Figure 2.9B Bedrock geology, central plain, with glacial sands.

South-Eastern Caledonian Province, or Leinster Batholith

The dominant feature here is the Leinster granite *batholith*, or large intrusion, which forms the Wicklow Mountains (see Figures 2.10A–C). The granite was intruded into Ordovician sediments, which have now mostly been eroded away except around the edges.

Figure 2.9C Bog of Allen, with Croghan Hill, Co. Offaly, in the distance. Sarah777. https://commons.wikimedia.org/wiki/File:Bog_of_Allen_-_Croghan_Hill.jpg.

Figure 2.10A Satellite image, southeastern Caledonian province.

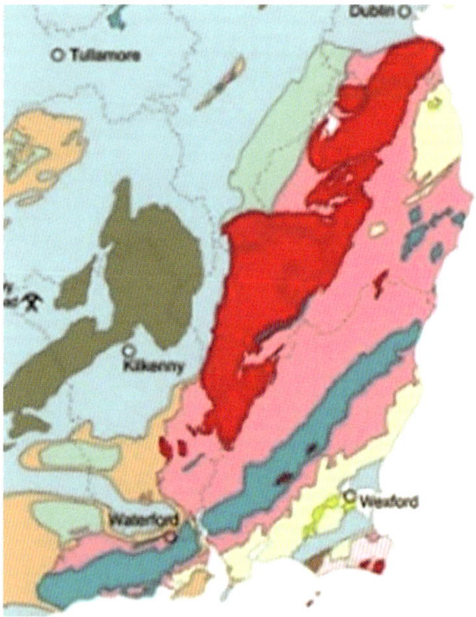

Figure 2.10B Bedrock geology, southeastern Caledonian province.

Figure 2.10C Lugnaquilla, Co. Wicklow, highest peak in the Leinster Granite batholith. Joe King. https://commons.wikimedia.org/wiki/File:Lugnaquilla_from_Glenmalure.jpg.

The Munster Ridge and Valley Province

The Munster Basin is characterized by parallel ridges of sandstone mountains of Devonian age, the *Old Red Sandstone,* forming mountain ranges such as MacGillycuddy's Reeks, separated by limestone basins. See Figures 2.11A–D, including an early example of animal tracks from around 400 mya on Valentia Island, Co. Kerry. The southwest coast has been flooded by the sea as a result of post-glacial uprise, and deep inlets known as *rias* have formed.

Figure 2.11A Satellite image, Munster ridge and valley province.

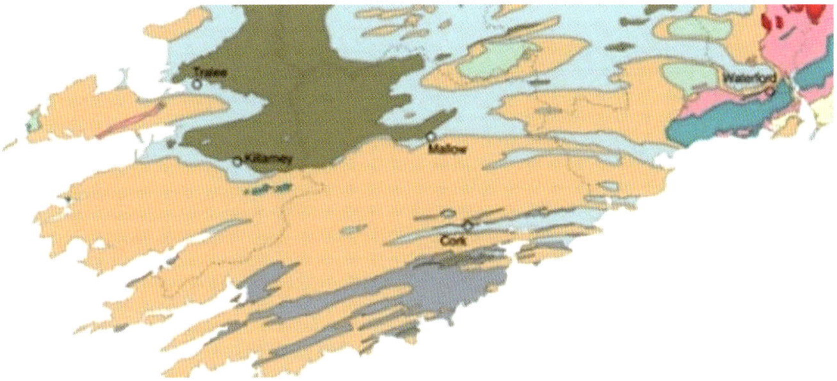

Figure 2.11B Bedrock geology; Munster ridge and valley province.

Figure 2.11C Carrauntoohil, highest peak in Macgillycuddy's Reeks. Enter. https://en.wikipedia.org/wiki/File:Carrauntoohil_2016.JPG.

Figure 2.11D An early example of land animal tracks from the Devonian era (~400 mya), Valentia Island, Co. Kerry. Photo courtesy of Mike Simms.

The Geological History of Ireland and the Opening and Closure of the Iapetus Ocean

That Ireland today exhibits such a wide variety of rock types of all geological ages is entirely due to a complex geological history that involves the movement of *crustal plates* over the surface of the globe. This movement is part of a process known as *Plate Tectonics*. By the mid-twentieth century, advances in the understanding of mantle and crustal processes and a deeper understanding of the extent of geological time had led to a revolution in the Earth sciences. The idea that the continents were not fixed in their positions, but were able to move, or *drift*, on the Earth's surface had been proposed by Alfred Wegener in the early part of the twentieth century. This proposal was largely rejected by the geological establishment as being impossible and the idea was abandoned except for a few recalcitrant supporters. One of these was the British geologist Arthur Holmes, who as well as supporting the idea of continental movement was also a pioneer in the

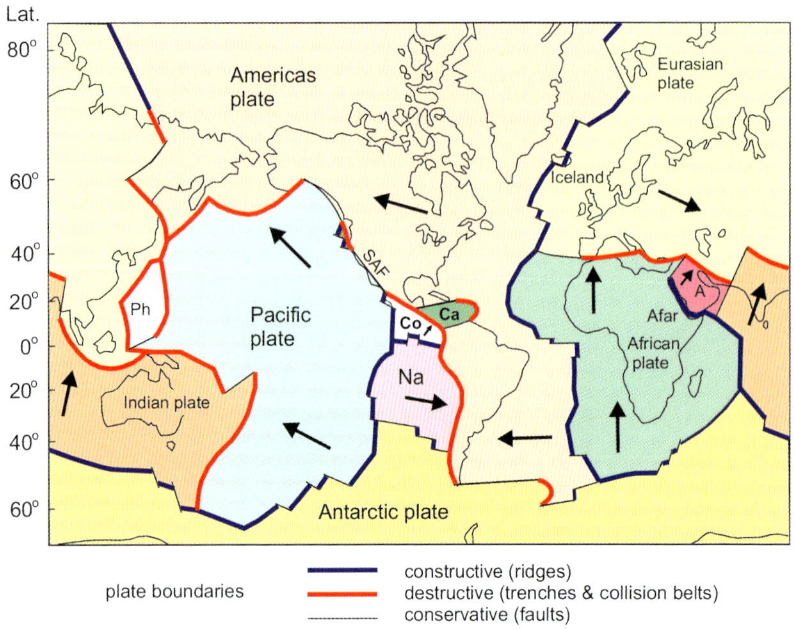

Figure 2.12 Distribution of tectonic plates on the Earth's surface.

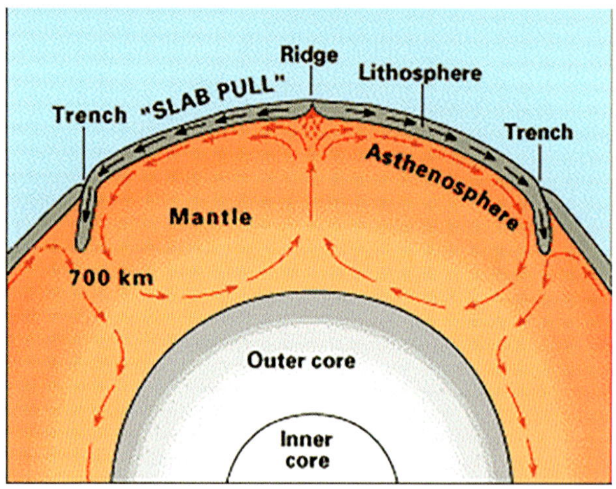

Figure 2.13 Convection cells in the Earth's Mantle (after United States Geological Survey).

development of radiometric dating, a way of measuring the absolute age of the Earth using radioactive isotopes that occur naturally in certain minerals. Holmes was able to demonstrate that the Earth was billions of years old, a much longer timescale than had previously been envisaged and which allowed the time required for the evolutionary processes envisaged by Charles Darwin and later workers on the origin and development of species. In addition, in the 1930s, Holmes proposed a convection cell mechanism whereby continents could split and separate to form oceans on the upward welling leg, with a corresponding downward welling leg destroying crust.

By the late 1950s and early 1960s, it had been established that the Earth consisted of a number of polygonal *plates* resembling the fragments of cracked shell on a hard-boiled egg (see Figure 2.12). Slow convection currents originating deep within the mantle bring hot material to the surface. This up-welling of relatively hot material is balanced by cooler, denser material sinking back into the mantle and thus completing the convection cycle. These rigid plates move on an underlying layer in the upper mantle, the *asthenosphere* (see Figure 2.13).

New crustal material is generated by separation at the mid-ocean ridges called constructive (divergent) margins. The oceans widen by a process called sea-floor spreading, with the new crust being generated from the underlying mantle. This expansion is balanced by loss of crust at destructive (convergent) margins, where the plates converge and collide. Crust is created at constructive plate margins such as the mid-Atlantic Ridge and destroyed in collisions zones such as the Andes in South America, where crust is pushed back into the mantle, or subducted, and is thus absorbed back into the mantle. These processes are summarized in Figure 2.14.

Where continental plates are forced together, the crust is lifted up to form major mountain ranges such as the Himalayas. They were formed when India began to move northwards, relatively rapidly, at about 15 cm per year, and collided with Asia around 60 mya (the

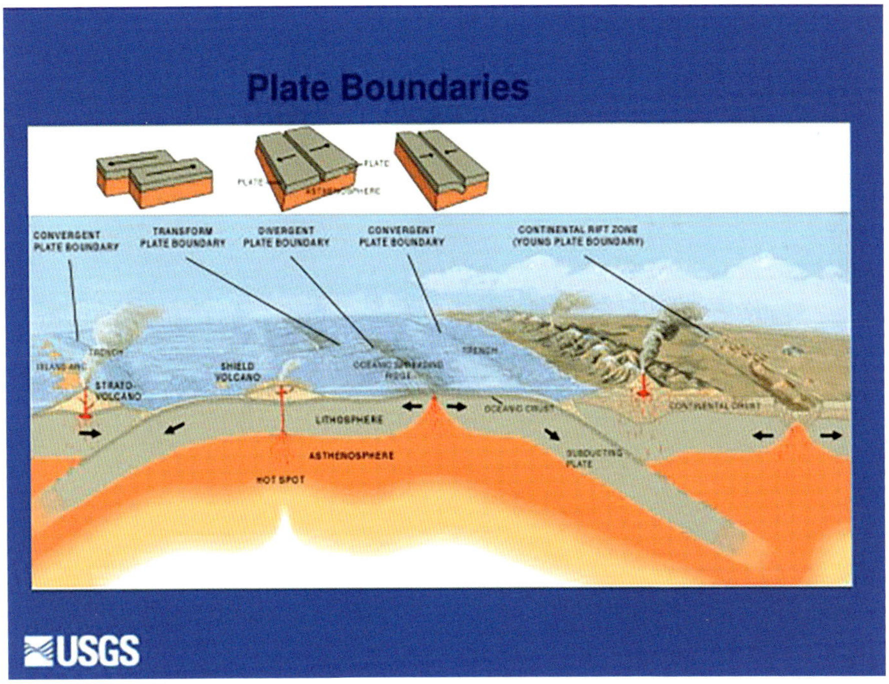

Figure 2.14 Types of plate boundaries (after United States Geological Survey).

ocean floor between India and Asia was systematically destroyed as the two continental plates converged and eventually collided). This movement continues with the result that the Tibetan Plateau to the north of India is still rising at the rate of a few millimetres a year.

The Wilson Cycle

In the 1960s, a Canadian geologist, John Tuzo Wilson, realized that if ocean basins are forming by the splitting apart of continents, then somewhere else on the Earth's surface other oceans must be closing, causing continents to collide and form mountain ranges – *the Wilson Cycle* (Figure 2.15). The Wilson Cycle operates over a timescale of a few hundred million years, and the extension of this cycle is to produce a *supercontinent* when all the continental areas of the Earth will amalgamate. There is evidence that supercontinents have formed and broken apart since at least ~3 billion years ago (gya), and the next one is predicted to form approximately 250 million years (ma) in the future.

The geological structure of Ireland can be examined in terms of the Wilson Cycle. The wide range of rock types found in Ireland is a result of the continuous amalgamation and rupture of continental masses during the working out of the plate tectonic processes as oceans opened and closed and supercontinents formed and dispersed. The geological structure of Ireland as seen today is largely based on events surrounding one particular mountain-building episode, the previously mentioned Caledonian Orogeny, which occurred around 490–390 mya. This orogeny affected the northern parts of Britain and Ireland and the Caledonian mountain range that resulted extended from Norway, across Scotland and Ireland into Newfoundland, eastern Canada and the Appalachians of the eastern USA (see Figure 2.16).

By around 1000 mya, the continents on the Earth's surface had merged to form the supercontinent Rodinia. From around 810 mya, Rodinia rifted and began to split apart to form two separate continents, *Laurentia* (North America plus Greenland) and *Gondwana* (South America plus Australasia and India) (see Figure 2.17A). The section of crust that now makes up Ireland as it exists today was divided between *Laurentia* and *Gondwana* that were separated by a widening ocean called *Iapetus*. The northern part of the island was part of Laurentia, while the southern part was on Gondwana.

The separation of Laurentia and Gondwana as this ocean expanded allowed the deposition of the sediments of the Dalradian Supergroup of rocks in northwestern Ireland in a zone that extends through Mayo, Donegal, Tyrone and northeast Antrim into the Scottish Highlands. The northern part of Ireland was located on Laurentia, while the southern part of

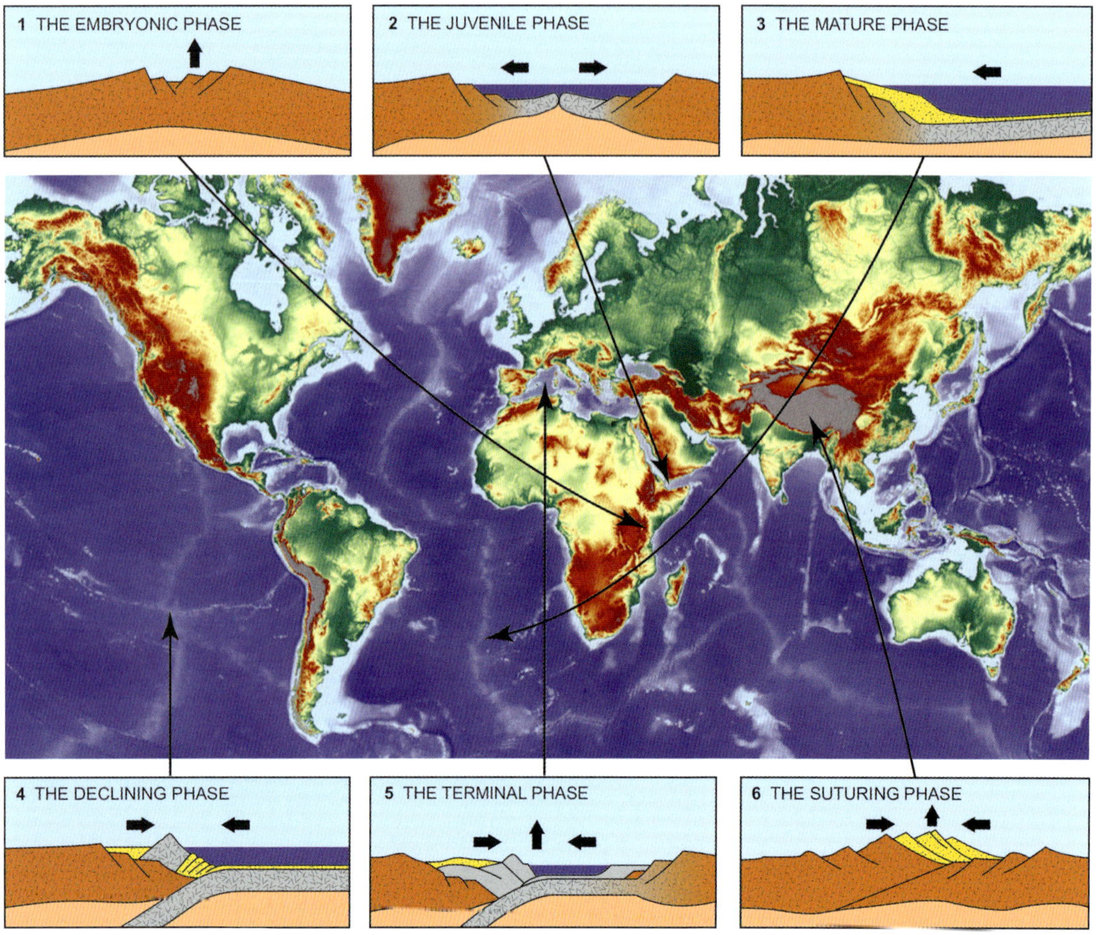

Figure 2.15 Stages of the Wilson Cycle, showing the formation, development and closure of an ocean basin (after Hannes Grobe).

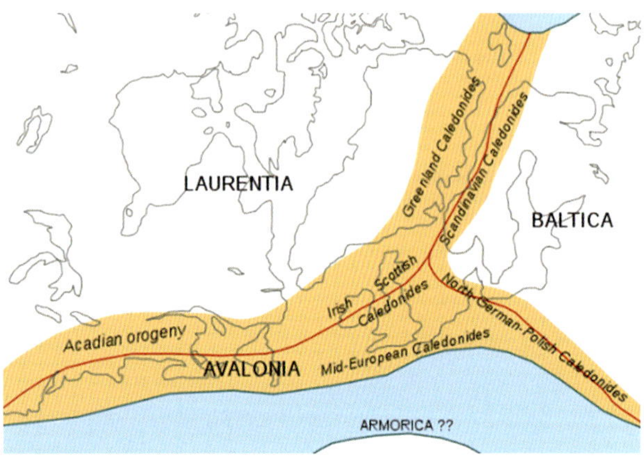

Figure 2.16 Reconstruction of America, Greenland and Europe following the formation of the Caledonides mountain chain, around 420 mya. Present-day coastlines are indicated in grey, for reference.

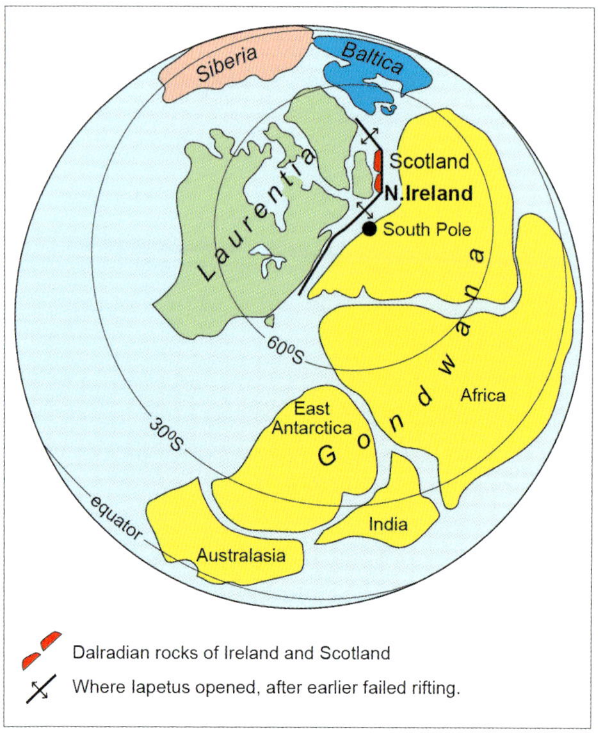

Figure 2.17A Breakup of supercontinent Rodinia, and the opening of the Iapetus Ocean *c.*600 mya. Courtesy of GSNI.

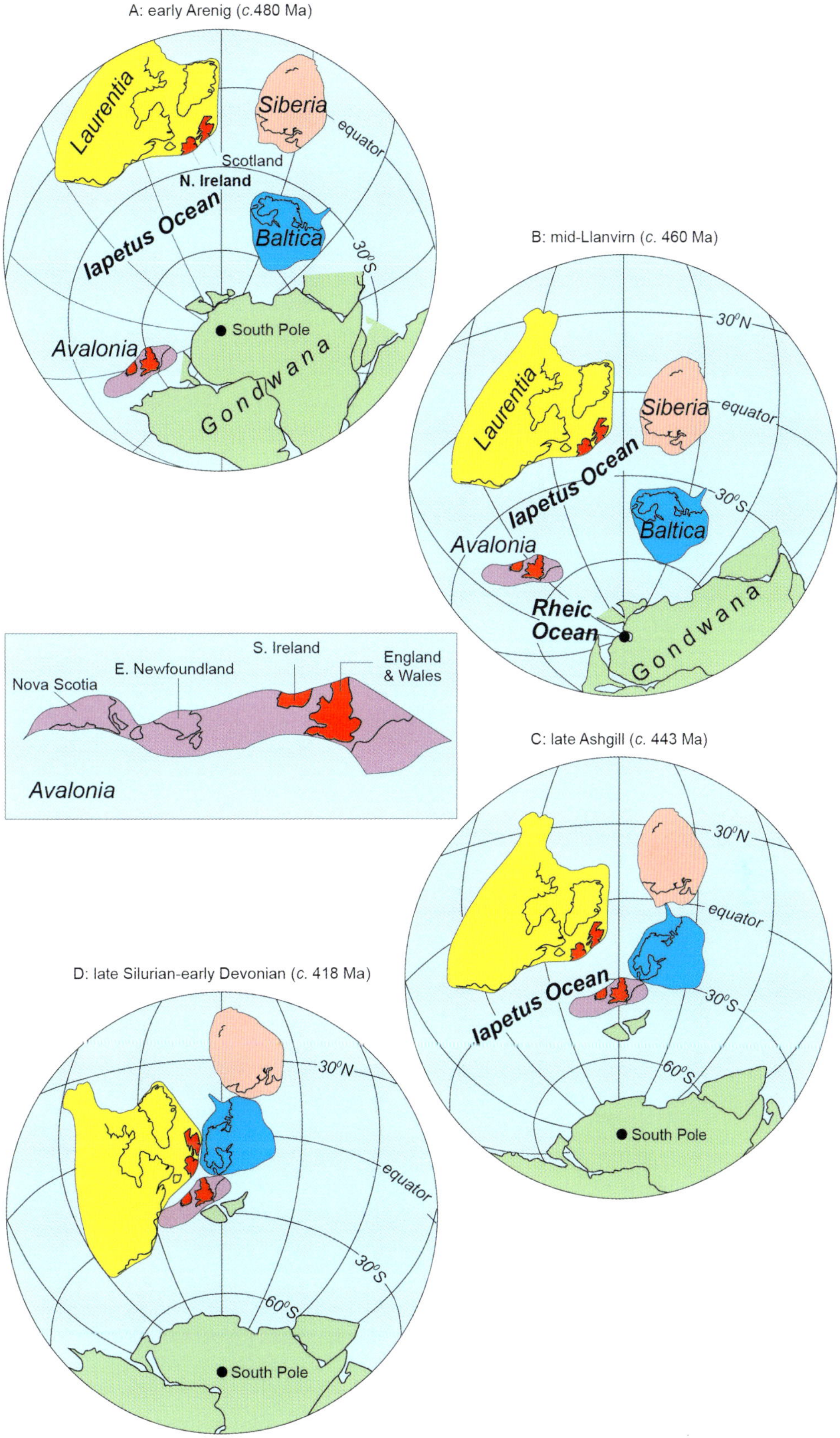

Figure 2.17B Sequence for the opening and closing of the Iapetus Ocean. Courtesy of GSNI.

Ireland was located on Avalonia, a volcanic arc on the northern margin of Gondwana, which eventually rifted off, becoming a drifting micro-continent (see Figure 2.17B).

A *terrane* in geology is a fragment of crustal material formed on, or broken off from, one tectonic plate and then added to – *accreted* or *sutured* – to another plate. The nature of subduction processes is such that fragments of lighter density continental crust are more buoyant than surrounding oceanic crust and are accreted on to adjacent continents when the oceanic crust was returned to the mantle. That fragment retains its own distinctive characteristics and geological history, and this may be very different to its new surrounding area. The *suture* or line of accretion between the two adjoining fragments can often be interpreted as a fault. Adjoining terranes can often display strikingly different geological environments and ages, with the implication they were once widely separated, possibly to the extent of 100s of kilometres.

The rocks of Precambrian and Palaeozoic age in Ireland can be regarded as a sequence of terranes where the boundaries are prominent fault systems, usually with an unknown amount of displacement along the fault. Around 470 mya (early Ordovician) the ocean had begun to close, and eventually the two continents collided (late Silurian) along a northeast–southwest axis that reflects the zone of collision. This zone – the *Iapetus Suture* can be traced from around the Shannon Estuary on the west coast to Clogher Head on the east coast (Figure 2.18A).

The suture separates Laurentian terranes to the northwest from *Avalonian* terranes to the southeast. *Avalonia* was a microcontinent in close proximity to Gondwana before closure of Iapetus and was swept up in the continental collision as Iapetus closed (see Figure 2.17B for the location of Avalonia as Iapetus opened and then closed).

Figure 2.18B shows the distribution of major rock units as they would have been 400 mya. Younger rocks are omitted. The terrane boundaries or sutures are mostly now concealed beneath younger rocks such as boulder clay and peat bog. The units did not all originate in their present positions, and some were formed thousands of kilometres from their neighbours. Assembly in their present position configuration was

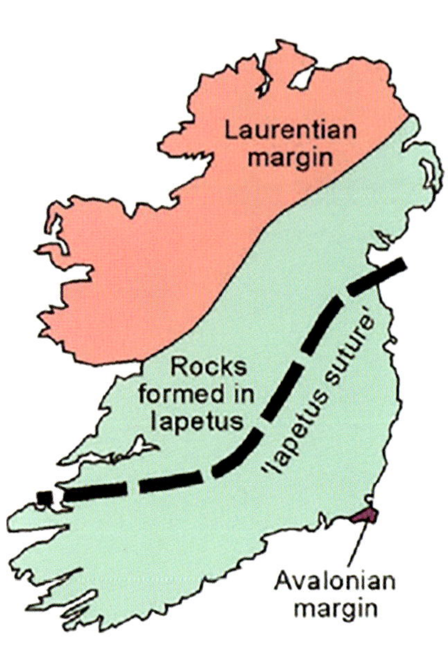

Figure 2.18A The approximate position of the Iapetus Suture. Courtesy of GSI.

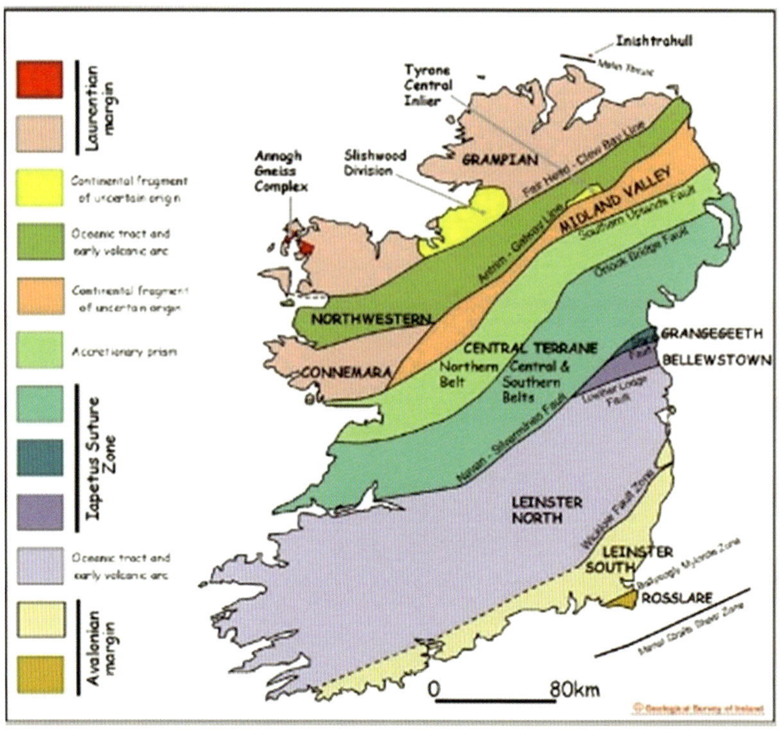

Figure 2.18B Terranes making up Ireland's basement rocks. Courtesy of GSI.

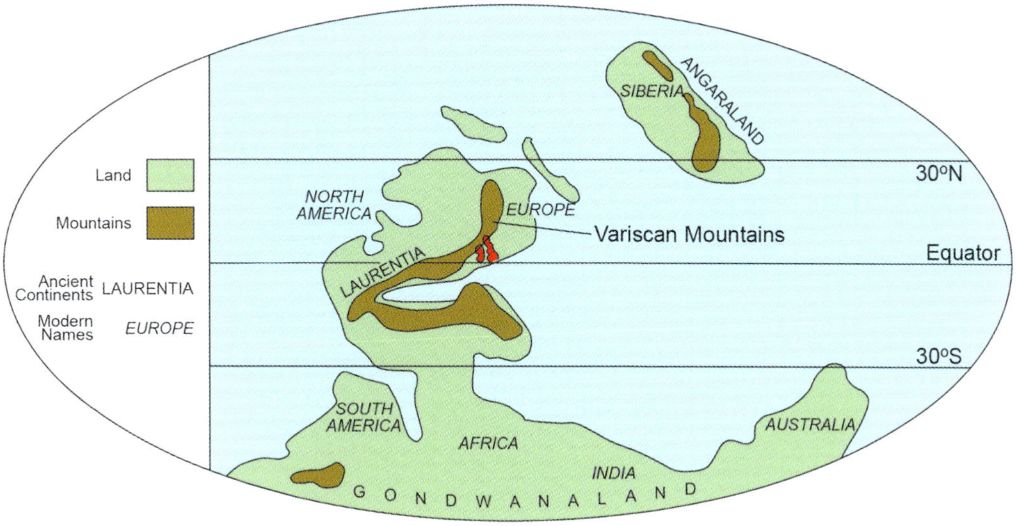

Figure 2.19 Position of the continents, late Carboniferous (~300 mya). Courtesy of GSNI.

driven by plate tectonic forces during closure of the ancient Iapetus Ocean. Between the ocean's opposite margins, present in the northwest and southeast are units comprising oceanic lithosphere, volcanic arcs, accretionary prisms and ocean basins.

One other consequence of this continental collision was to lay the geological foundations of Britain and Ireland as they exist today. The closure juxtaposed the northern parts of Britain and Ireland, which were part of Laurentia, and the southern parts which were part of Avalonia. The final stages of the closure of the Iapetus Ocean were marked by voluminous granite intrusions during late Silurian and early Devonian times (420–390 mya). The granites of Co. Donegal in northwest Ireland were emplaced during this phase of crustal development. Sedimentary basins formed and with the high sediment supply from the Caledonian mountains to the north, thick deposits of sandstones and conglomerates developed, the *Old Red Sandstone*. These are especially well seen in the Munster region of southwest Ireland.

Figure 2.20 Folded sedimentary rocks of Carboniferous age at Loughshinny, Co. Dublin (folding occurred during the Variscan orogeny; the height of the folds is about 6 m). https://commons.wikimedia.org/wiki/File:Loughshinny_folds.jpg Siim Sepp (Sandatlas), CC BY-SA 3.0 https://creativecommons.org/licenses/by-sa/3.0, via Wikimedia Commons.

Later, the area now recognizable as Ireland moved northwards to reach the Equator by around 300–350 mya (see Figure 2.19). This period was the Carboniferous, with marine conditions expanding over Ireland from the south. Initial deposits were sandstones, shales and mudstones in shallow seas followed by extensive limestone deposits that formed in a warm tropical sea. Sea level changes followed, forming swampy conditions, with rapid accumulation of vegetation that formed coal deposits in areas such as Co. Roscommon at Arigna or in the Leinster coalfield. There is a noticeable difference in the relative scarcity of coal deposits in Ireland when compared with the substantial amounts found in England, Scotland and Wales. This discrepancy is a consequence of post-Carboniferous erosion and had major implications for the growth of industrialization in Ireland in the mid-nineteenth century, which will be discussed in a later chapter.

The continental collisions producing this supercontinent are known as the *Variscan* or *Hercynian* orogeny, and with pressure coming from the south produced strong folding in the southern part of Ireland (see, for example, Lough Shinny, in Figure 2.20). In addition, the east–west trending mountains of south Munster have a very different trend from the earlier Caledonian mountains in Ireland, which trend northeast–southwest.

Gondwanaland continued to move northwards in collision with Laurentia, and the continents had converged by the late Palaeozoic (around 250 mya)

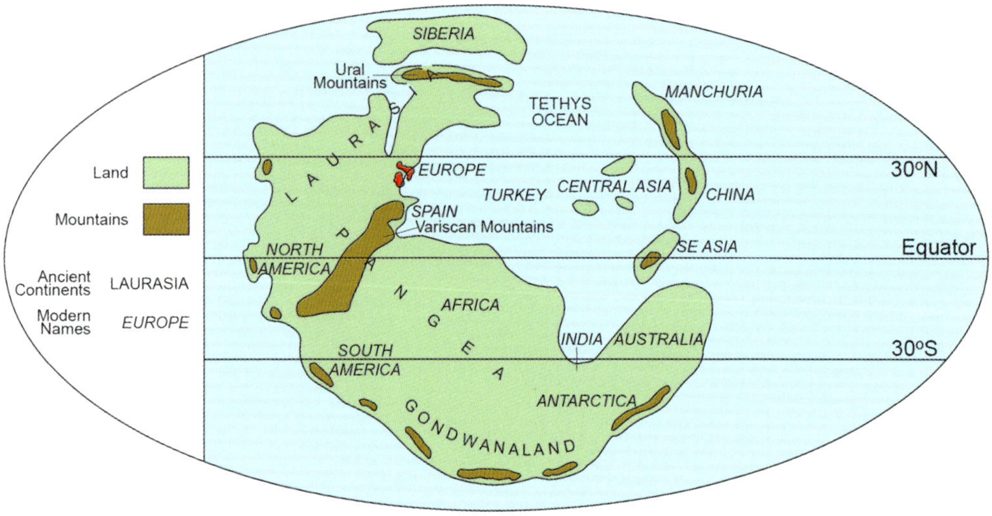

Figure 2.21 Position of the continents, early Triassic (~250 mya). Courtesy of GSNI.

to form the *supercontinent*, Pangea. The final assembly of the components of Pangea had taken place by early Permian times, around 2993–273 mya (Figure 2.21). During the Triassic and Permian periods, Ireland was within the continental area of Pangea and in a largely arid climate regime. Fine-grained desert sandstones and coarser conglomerate beds were typical deposits. Successive cycles of shallow seas coupled with high evaporation rates led to deposits of *evaporites* such as *gypsum* at Kingscourt in Co. Cavan, and *halite* or *rock salt* in Co. Antrim. The supercontinent existed until early Jurassic times when it began to break up about 200 mya. During the succeeding Cretaceous Period, most of Ireland was covered by a shallow sea with deposits of white limestone or chalk and abundant marine fossils. Rocks of Cretaceous age are only found in Northern Ireland, probably having been eroded from elsewhere on the island.

Eventually the distribution of continents resembled the pattern we are familiar with today, with the Indian and Atlantic Oceans forming and widening by Palaeocene (formerly Tertiary) times, around 60 mya (see Figure 2.22). A consequence of the opening and spread of the Atlantic Ocean was that Ireland was subjected to a prolonged period of magma production associated with a *mantle plume* that affected Ireland, Scotland and Greenland, and eventually formed Iceland on the mid-Atlantic Ridge.

The Antrim Lava Group in Co. Antrim and Co. Londonderry, which includes the columnar basalts of the Giant's Causeway, is the legacy of this magmatic episode, with the eruption of extensive lavas and the intrusion of associated dykes and sills some 60 mya.

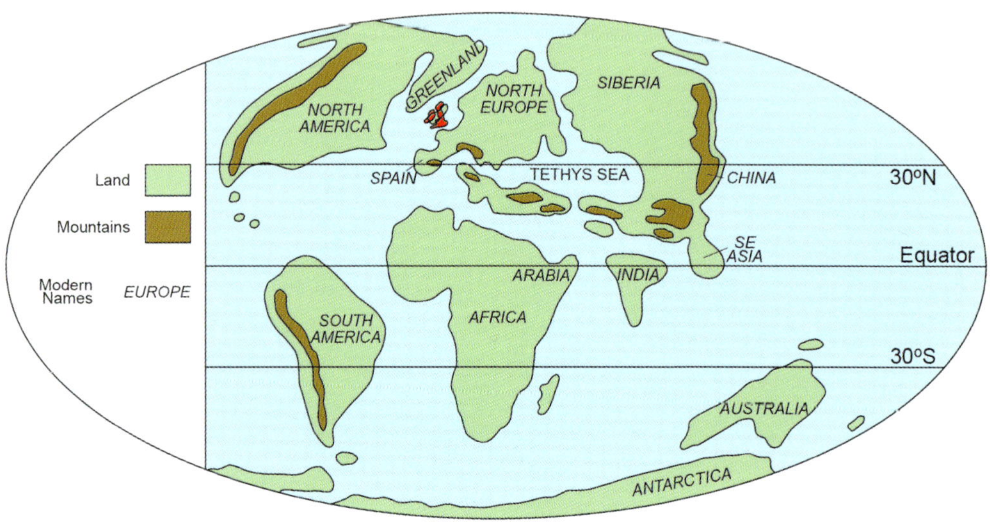

Figure 2.22 Position of the continents in Palaeocene times (~60 mya). Courtesy of GSNI.

The Immediate Post-Glacial Landscape and the Stages in Its Subsequent Modification

Ireland's landscape was formed during the *Quaternary* era, which started about 2.5 mya. Its unconsolidated deposits, referred to as superficial, cover most of the older rocks and contribute to modern soil formation.

Table 2.1 Quaternary period subdivisions by epoch and start date

System/period	Series/epoch	Start/date
Quaternary	Holocene	11.7 kya
	Pleistocene	2.6 mya

The final glacial period of the Quaternary has had a profound influence on the landscape we see in Ireland today. The Pleistocene epoch is the older of the two epochs making up the Quaternary, the younger being the Holocene. The Pleistocene began 2.6 mya and ushered in a period of global cooling resulting in vast accumulations of snow and ice in the Northern Hemisphere and giving a series of glacial and interglacial cycles. The Holocene covers the last 11.7 kyr and represents the youngest interglacial interval of the Quaternary period. The Holocene is significant because it includes the evolution and development of *Homo sapiens*, with the attendant impacts on the landscape and environment (Table 2.1).

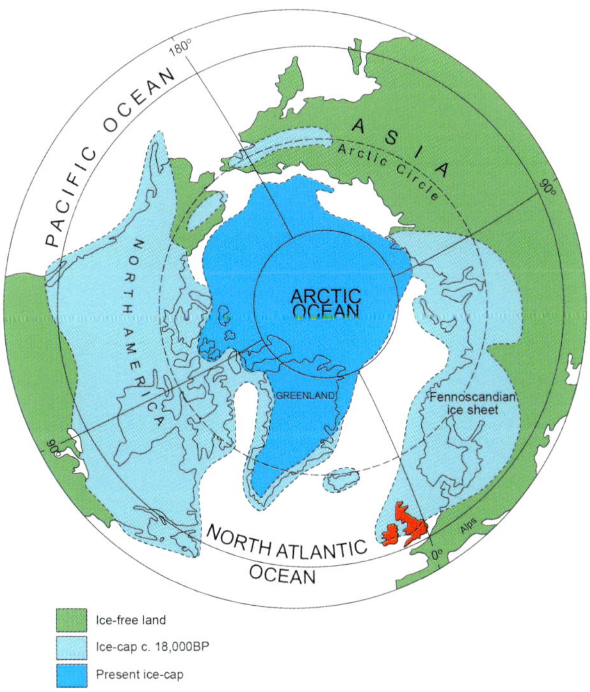

Figure 2.23 Distribution of ice sheets during late Pleistocene and at present in the Northern Hemisphere. Courtesy of GSNI.

While the bedrock geology comprises a wide range of rock types and ages, the present landscape is mostly the result of more recent geological events and processes. Predominant among these were glacial processes of erosion and deposition acting as the ice melted and the landscape emerged from its icy carapace some 12,000 years ago. Each glacial cycle resulted in successions of tills, sands, gravels and clays deposited in the distinct phases of each cycle. Figure 2.23 shows the distribution of ice sheets during the late Pleistocene and the current extent of ice-caps in the Northern Hemisphere.

The landscape we are familiar with today derived from this post-glacial landscape and its subsequent modification. Changes were accelerated after the arrival of the first settlers around 9000 BP, who made the journey by sea with a landfall probably on the northeast coast of the island.

At the height of the glaciation period, around 20,000 years ago, when virtually all of Ireland was ice-covered, and with sea levels lower than at present, a land bridge connected Britain and Ireland. As the ice sheets retreated, the land bridge gradually submerged, isolating Ireland from both Britain and Europe. On the lowland areas, an environment known as tundra developed, characterized by large stretches of bare ground, or rock with patchy covering of low vegetation, such as mosses, lichens and dwarf shrubs, and where the sub-soil may be permanently frozen, a condition known as *permafrost*.

The *drumlin belt* referred to earlier in the chapter represents a distinctive landscape over much of a zone stretching from Co. Down on the northeast coast across to the west coast around Counties Donegal and Sligo. The typical low hills, and the intervening hollows with poor drainage, led to a tendency to small fields, with the farmhouses concentrated on the higher ground. The characteristic shape of drumlins is best seen in areas where they have been inundated by post-glacial sea level rises, for example Strangford Lough (Figure 2.24) or Clew Bay in Co. Mayo (Figure 2.7C).

Figure 2.25, *Landscape, County Down* by Colleen Frew, is a representation of a typical hilltop farmhouse

Figure 2.24 Drowned drumlins, Strangford Lough, Co. Down. NIEA.

Figure 2.25 *Landscape, County Down*, by Colleen Frew. Courtesy of the artist.

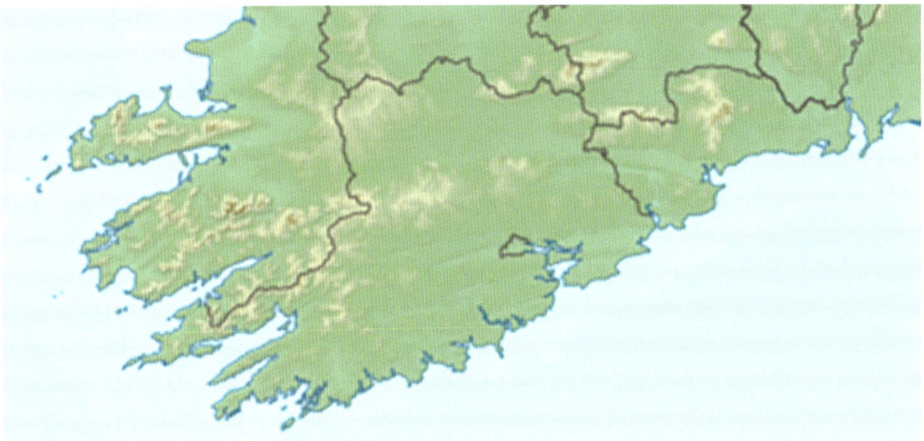

Figure 2.26 Rias and ria coastlines. Similar coastal features exist in other parts of the world, including northwest Spain and the southwest coast of Australia.

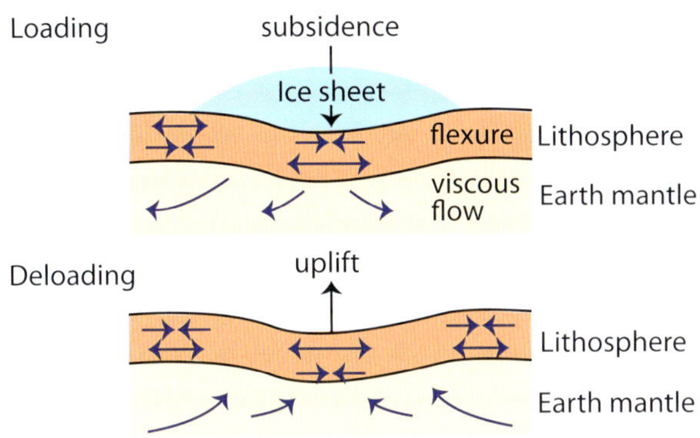

Figure 2.27 Isostatic readjustment of the crust after melting of ice sheet, causing a fall in sea level.

Figure 2.28 Former sea stacks on raised beach, Ballintoy, north Antrim coast.

in drumlin country, with a small lake in the inter-drumlin hollow.

An important cultural response to life in the drumlin belt were the writings of John McGahern (1934–2006), who lived and farmed in Co. Leitrim, in the middle of the drumlin belt. He wrote extensively about the economic and social realities of life on a small farm in the 1930s and 1940s. He dealt with the conflicts between the traditional mindsets of the older generation and a younger population anxious for new horizons and opportunities in an Ireland on the cusp of major changes in society in the second half of the twentieth century.

Glaciation and Sea Level Changes

Changes in sea level are *eustatic* or *isostatic*. Eustatic changes are global sea level changes related to changes in the volume of water in the ocean. A rise or fall in sea level can be caused by changes in the amount of glacial ice on land, or to the thermal expansion of water with increasing temperature, or on a longer timescale by plate tectonic forces. For example, if an ocean basin is widening, the volume of water contained will change and thus sea level will be affected.

Over the period covered by the Ice Age, global sea level was, on average, some 130 metres lower than it is today due to the enormous amounts of seawater stored in ice caps and glaciers. Changes due to the melting of ice caps occur relatively rapidly, so some 20,000 years BP there was a global rise in sea level as the ice sheets began to melt. In southwest Ireland, in the Counties of Cork and Kerry, a series of long narrow inlets were formed by the submergence of pre-existing river valleys as sea level rose. Bantry Bay and the River Bandon are good examples of the landform termed *rias*, formed by the submergence of a pre-existing river valley (see Figure 2.26).

Isostatic sea level changes are local changes caused by

subsidence or *uplift* related to processes such as changes in ice cap thickness on land, or formation or erosion of mountain ranges. Parts of the Northern Hemisphere were covered by ice, up to several kilometres thick in places. This load had originally caused the crust to *downwarp* by deforming the underlying mantle due to a process called *isostatic readjustment* (Figure 2.27). When the ice melts, the crust is unloaded and the mantle flows back in, raising the crust, and so the sea level falls. This is known as an *isostatic sea level* and is a relative change since the land is rising. Isostatic sea level change is the result of an increase or decrease in the height of the land. When the height of the land increases, the sea level falls, and when the height of the land decreases, the sea level rises. Isostatic change is a local sea level change whereas eustatic change is a global sea level change. Isostatic changes are slower than the relatively rapid eustatic changes. The effect of this rise in the north of Ireland is a net land rise of about 22 metres in the last 16,000 years and the formation of raised beaches on the north coast, with features such as former sea stacks above the current high water mark (Figure 2.28).

A result of these post-glacial changes in sea level, which are ongoing because of the slow rate of change of the mantle adjustment, is that Ireland is tilting to the south, with the north of the island rising. The land bridge between the north of Ireland and Scotland ceased to exist around 11,000 years BP, and the bridge between the southeast of Ireland and southwest England disappeared about 14,000 years BP.

Following the retreat of the ice and the gradual warming of the climate, the landscape underwent a series of changes, accelerated by the arrival of the earliest settlers around 9,000 years BP. Important landscape developments in this phase were the growth of peat bog; the development of agriculture allowing the change from a hunter–gatherer society to an agrarian way of life; and the parallel development of megalithic monuments in a ritual landscape. The significance of peat bogs in the culture of Ireland will be dealt with at length in a later chapter.

3

Mapping the Landscape

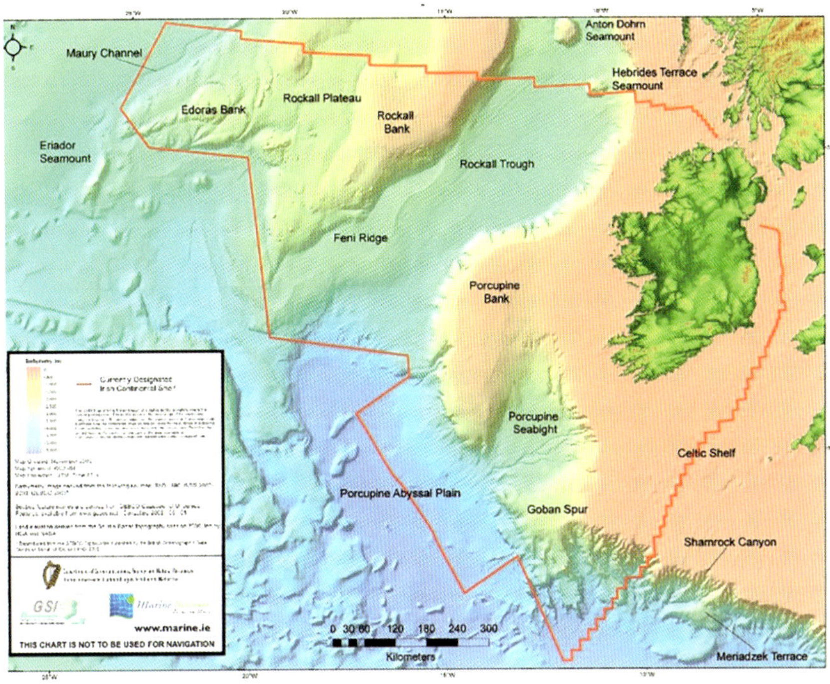

Figure 3.1 Irish territorial waters (continental shelf). Courtesy of GSI.

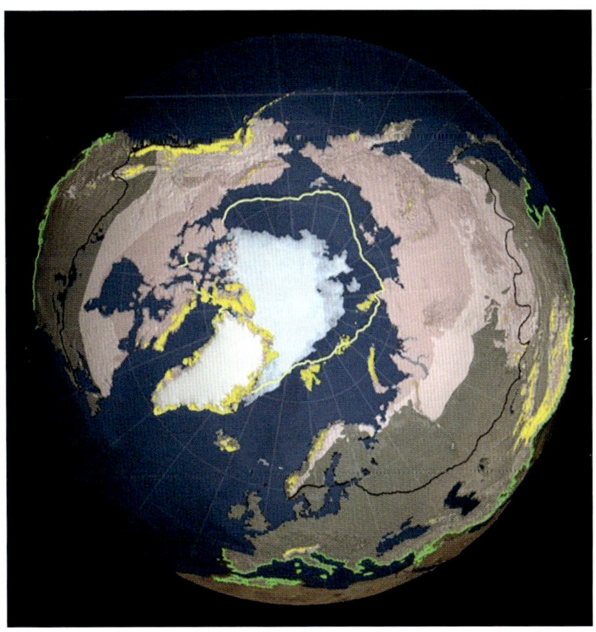

Figure 3.2 Projection showing position of Ireland in the Northern Hemisphere. Image courtesy of NASA.

The Real Map of Ireland shows Ireland's marine territory of over 90 million hectares (220 million acres or 900,000 square kilometres), which is ten times the size of the island of Ireland. The designated Irish Continental Shelf, represented by the red line on the map, shows Ireland's current territorial waters, which extend out across the Atlantic Ocean and cover parts of the Irish and Celtic Seas (Figure 3.1). This area also includes one of the largest marine Exclusive Economic Zones in the European Union. This map represents the cutting edge of modern map-making technology, incorporating underwater surveying techniques to map the extent of the continental shelf around Ireland, and sophisticated satellite data collection to provide an accurate map of the terrestrial features. Figure 3.2 shows the position of Ireland in the Northern Hemisphere, on the eastern fringes of the Atlantic Ocean.

Figure 3.3 Position of Ireland in relation to western Europe. Image courtesy of NASA.

In closer focus, Figure 3.3 shows Ireland as part of an archipelago, a group of islands situated on the western fringes of continental Europe. This grouping is sometimes referred to as the British Isles, although this term is not universally accepted. Further images from space allow Ireland to be placed in a global context. Ireland's relationships with its near neighbours, Britain and continental Europe, underwent profound changes after the discovery of the so-called 'New World' after 1492 CE,[3] with the voyages of Christopher Columbus and those who followed him across the Atlantic.

During the period of western history that was centred on the Mediterranean Sea and dominated by the early civilizations of Egypt, Greece, Persia and Rome, Ireland was on the western periphery. *Thule* (Latin) is the location farthest north mentioned in ancient Greek and Roman literature, and in classical and medieval literature, *Ultima Thule* became a metaphor for any distant region considered to be beyond the borders of the known world. This would have included Ireland, northern Scotland, the Orkney and Shetland Islands, Iceland and Greenland. Early voyagers from the Mediterranean were the Phoenicians, who are known to have travelled to Cornwall in southwest England in their search for tin ore, and to the Baltic region to obtain amber. It is likely that these intrepid sailors travelled further west and northwest to encounter Ireland and Scotland.

3 CE (Common Era), used here instead of AD (Anno Domini, in the year of our Lord). BCE (before the Common Era) is used instead of BC (before Christ).

Following the colonization of the Americas by European powers and the opening of transatlantic trade routes, the significance of Ireland – on the fringes of continental Europe and with its back to the Atlantic Ocean – changed. Rather than one of the remote regions in the context of a civilization based around the Mediterranean, Ireland now found itself facing this new world, with all the sociological and economic implications that this involved. A comprehensive treatment of this change in the geographical context of Ireland is to be found in Barry Cunliffe's *Facing the Ocean* (2001).

The period after the opening of north Atlantic trade routes coincided with the growth of Ireland as a trading and manufacturing nation, with the concurrent expansion of its internal communication networks as already described. As well as reorienting itself in the context of the outside world, Ireland needed a formalization of its internal features. As in Scotland in the eighteenth century, internal strife alerted the government, not only to a necessity to improve the road network, but also to the lack of a proper map of the jurisdiction. This led eventually to the formation of the Ordnance Survey in Britain in the early part of the nineteenth century, following a military survey some thirty years earlier. The term 'ordnance' can mean artillery or mounted guns and also a branch of government service dealing with military stores and material. That the Ordnance Survey was responsible for mapping Ireland had political and social implications for many in the population.

Historical Aspects of Map-Making

The science of cartography or map-making is nowadays viewed as an exercise in precision. The images discussed above come after a long history of representing the topography of our 3-dimensional surroundings on to a 2-dimensional sheet of paper. Our expectations are high that modern maps, whether in analogue or digital form, and at whatever scale, will be an accurate representation of a given place or region. This perception, however, is a product of the scientific revolution(s) that arose from the Middle Ages onwards. Previous to this, the making of maps was an endeavour that involved much more than the placing of a geometric grid upon space. The production of a map involved the history as well as the geography of the location, included mythology and folklore that invoked memories and combined supposition with superstition. Early maps had little sense of exactitude in direction or scale. Rather, they could be thought of as narratives, often told by individual travellers but also by cultural groups or tribes as they moved from one region to another.

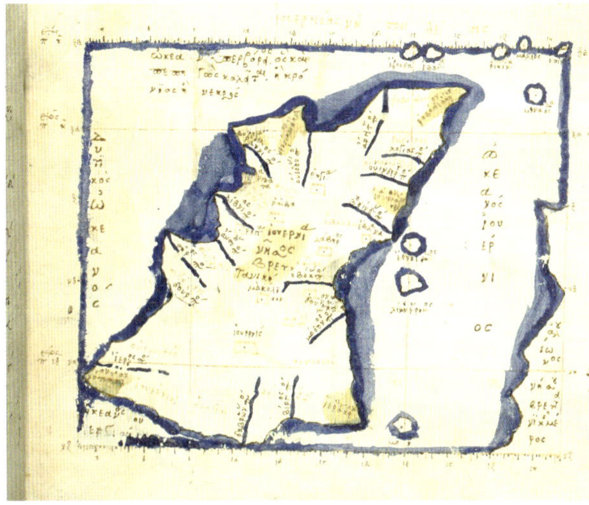

Figure 3.4 Map of Ireland after Ptolemy's *Geographica*, drawn around 1400CE. https://commons.wikimedia.org/wiki/File:Prima_Europe_tabula.jpg.

Early Attempts at Mapping Ireland

The earliest known attempt to produce a map of the whole of Ireland was by the Greek geographer and cartographer, Claudius Ptolemy, in 140 CE. Ptolemy produced an interpretation of the world based on observations and recordings from sailors from various parts of the Mediterranean who ventured out into the northern Atlantic for trade. For this reason, the most accurate parts of the constructed map are the coastal regions (Figure 3.4).

Figure 3.5 Map of Ireland, by Baptista Boazio, 1599. https://commons.wikimedia.org/wiki/File:Irlandiæ_accurata_descriptio,_by_Baptista_Boazio_(thumb).jpg.

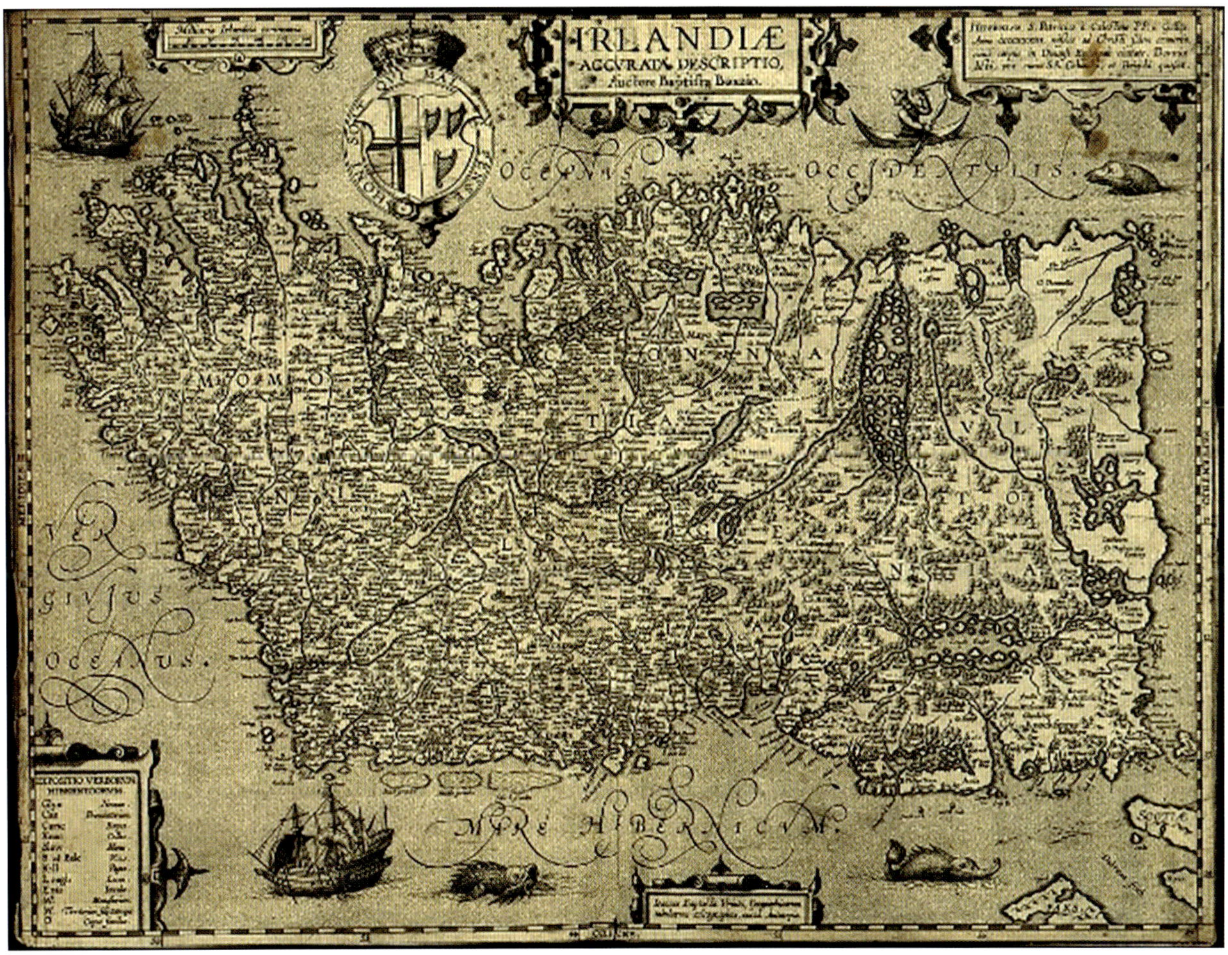

By the end of the sixteenth century, matters had improved considerably, and a version by Baptista Boazio (1599), with north to the right, shows not only the four provincial boundaries, but the counties are delineated by dotted lines. Additional details include pictorial representations of forests, topographical features and the locations of major towns and cities (Figure 3.5).

The Ordnance Survey in Great Britain: Making a Modern Topographical Map

After Boazio's map in the sixteenth century, the next important step in cartography was the application of scientific principles to ensure more precision in the making of maps. In Britain and Ireland, this meant the formation of a national survey (Figures 3.6A–B).

Triangulation is a surveying method that measures the angles in a triangle formed by three survey control points. Using trigonometry and the measured length of just one side, the other distances in the triangle are calculated. Measurement is taken to a triangulation pillar or trig point (see Figure 3.6B). It is usually a concrete pillar, tapering towards a flat top that has a metal plate to hold the theodolite, the instrument used to measure the angle and distance to the next triangulation point. 'The Principal Triangulation of Britain' was the first trigonometric survey of the whole of Great Britain and Ireland and was carried out between 1791 and 1853 (see Figure 3.7).

Following the 1745 Jacobite Rebellion in Britain, the British Government realized that not only were the roads in the Scottish Highlands inadequate for military action against the rebels, but also that there were no adequate maps for the area. So, as well as General Wade's road building programme for the Highlands (see Chapter 9), between 1747 and 1755, a survey of Scotland was undertaken at a scale of one inch to 1,000 yards (1:36,000). The survey was

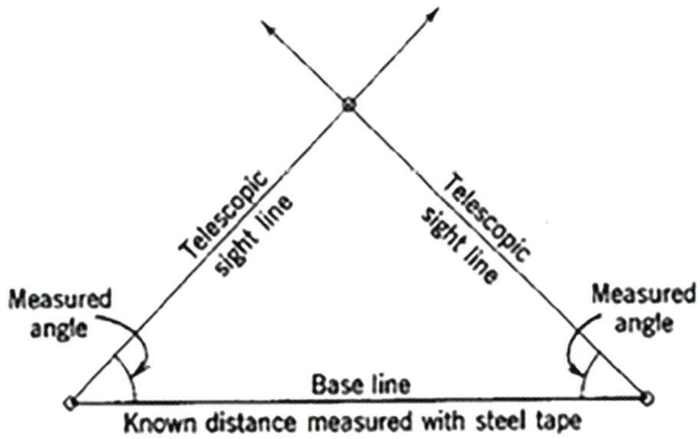

Figure 3.6A Principle of triangulation.

Figure 3.6B Cloghan Trig Point, Co. Galway, by Jason Keane. https://commons.wikimedia.org/wiki/File:Cloghan_Trig_Point_-_geograph.org.uk_-_7736201.jpg.

Figure 3.7 The nineteenth-century triangulation mesh over Britain and Ireland. Peter Mercator. https://commons.wikimedia.org/wiki/File:Clarke_principal_triangulation_of_Britain_1860.jpg.

carried out by Colonel David Watson, an Army officer attached to the Engineers of the Board of Ordnance. This was the government body responsible for, among other things, acting as custodian of the lands, depots and forts required for the defence of the realm and supplying munitions and equipment to both the Army and the Navy. In addition, the board also maintained and directed the Artillery and Engineer corps. As mentioned earlier, the term ordnance can mean both a mounted gun or a branch of the government dealing with military stores and material. Thus, the process of the formal surveying in Britain was undertaken as a military exercise/function, and this military involvement was retained in Britain until 1870, when it passed from the War Office to civilian control. The formal start of the Ordnance Survey in Britain is generally recognized as June 1791, when the third Duke of Richmond, in his role as Master-General of the Ordnance, authorized the purchase of a new state of the art theodolite. Staff were recruited to carry out the work, and by 1823 it had covered much of Britain.

The Ordnance Survey in Ireland

As in Britain, the formal beginning of survey work in Ireland was preceded by a military survey, carried out in this case by General Charles Vallency, between 1776 and 1790. This was a survey at scales of 1:20,160 and 1:40,320 and covered most of Ireland south of a line from Arklow to Limerick. Vallency's survey took place when Ireland was still technically an independent country, with its own parliament, but following the 1800 Act of Union, it became part of the United Kingdom of Great Britain and Ireland. This Act of Union followed quickly after the 1798 rebellion in Ireland, so the mapping of Ireland by the Ordnance Survey centred in Britain was a sensitive matter. Initially there was little enthusiasm from the Ordnance Survey for an Irish mapping project, but in the mid-1810s, for reasons of taxation rather than any military consideration, the mapping of Ireland became a relevant issue.

The term 'cess' in Ireland was a tax to fund facilities such as schools and prisons, to repair roads and bridges and to pay the salaries of local officials. The word is derived from 'assess' and could be considered as the equivalent of the modern-day rates or council tax on property. The county cess valuations were based on townlands, local divisions of land smaller than parishes. A 'townland' (Irish: *baile fearainn*) is a division of land of Gaelic origin that pre-dates the Norman invasion. Townland areas ranged from 300–600 acres (120–240 hectares). Although the names and rough outlines were generally known, the precise boundaries and areas were poorly marked. As the nineteenth century progressed, this lack of information began to worry those responsible for tax collection.

In 1824, Thomas Colby became Superintendent of the Ordnance Survey in Britain, having succeeded General William Mudge in July 1820. A Select Committee of the House of Commons had recommended a 1:10,560 (6 inches to the mile) survey of townland boundaries and of such details as could be shown on a 1:63,360 (1 inch to the mile) map. Instead, Colby ordered that all details should be mapped on the 1:10,560 scale (6 inches to the mile), with as much accuracy as this scale permitted, except for field boundaries, and after a few years they too were being mapped. Following some years when there was a necessity to repeat much unsatisfactory work, by the late 1830s, things were progressing smoothly. Most of the workers on the Irish survey, over 2,000 at its peak, were locally recruited civilians, but they were under supervision by the military officers and men of the Royal Engineers and the Royal Sappers and Miners, with three companies of sappers and miners trained in surveying duties. See Figure 3.8.

In May 1833, the first county, Londonderry, was completed. The rest of the country followed, and the map was completed in 1847 (Figure 3.9). Many of the problems associated with a military-controlled land survey in Ireland in the early part of the nineteenth century are dealt with in the play *Translations* (1980), by Brian Friel, which deals with issues of language, Irish history and cultural imperialism. It is set in a small village in Co. Donegal, in 1833, an isolated Irish-speaking community whose residents have had little contact with the outside, English-speaking world. This all changes when two British Army Royal Engineers arrive as part of the first Ordnance Survey. The remit of the soldiers is to make maps of the area, which includes translating the Irish names of locations and landmarks into English. While this may appear to

Figure 3.8 Sappers and miners at work. http://www.militarysurvey.org.uk/.

be a relatively straightforward administrative process by a government agency, such is the degree of culture, history and tradition embedded in the Irish language, that there are many significant ramifications. Themes dealt with in the play include language, education, colonization and the usurping of one cultural identity with another. In many respects, such themes are still extant in modern Ireland, with concerns being expressed about the apparent demise of the Irish language. Along with this, there are other perceived cultural misappropriations taking place in the context of globalization and a multicultural society, where many of the traditional influences and constraints have been abandoned.

The Geological Survey in Ireland

Unlike the situation in Britain, in Ireland, Colby instructed his officers to keep journals and produce memoirs covering a wide range of environmental, sociological and historical information. In 1824, he appointed Joseph Ellison Portlock to oversee the topographical mapping of Ireland for the Ordnance Survey. Portlock was a geologist and officer in the Royal Engineers. The information gathered during the survey of the parish of Templemore in Co. Londonderry was used by Portlock to produce a geological memoir and map. So, in 1843, *Report on the Geology of the County of Londonderry and parts of Tyrone and Fermanagh, Examined and Described Under the Authority of the Master General and Board of Ordnance* was published. The completion of Portlock's memoir and geological map was of great significance to the future surveying of the geology of Ireland, representing as it does the

Figure 3.9 Map of Co. Londonderry, 1837. S. Lewis & Co. https://commons.wikimedia.org/wiki/File:1837Londonderry.jpg.

Figure 3.10 Portrait of Sir Richard Griffith (1784–1878), holding a geological map of Ireland. Stephen Catterson Smith. https://commons.wikimedia.org/wiki/File:Portrait_of_Sir_Richard_John_Griffith,_1st_Bt._(by_Stephen_Catterson_Smith).jpg.

precursor to the formation of the Geological Survey of Ireland, in 1845.

At around the same time as Colby was developing a geological survey as part of the remit of the Ordnance Survey, Sir Richard Griffith was also carrying out a geological survey from the Valuation Office in Dublin. He had been running an unofficial geological survey within the Valuation Office that had produced the first geological map of Ireland at the scale of 1:254,440, or four miles to the inch, in 1839. Sir Richard Griffith (1784–1878) was an Irish geologist and mining engineer who was chairman of the Board of Works of Ireland (Figure 3.10). In 1825, he was appointed to carry out a boundary survey of the island and subsequently was made Commissioner of Valuation, producing *Griffith's Valuation*, used for various local and public assessments. He published several improved versions of his geological map, in 1839 and 1855, and for his contribution to geological science he was awarded the Wollaston Medal of the Geological Society of London, its highest award.

After a period of intrigue, and wheeling and dealing among the principal players in both Ireland and Britain, in 1845, Sir Henry Thomas De La Beche became the founder and first Director of the *civilian* Geological Survey of the United Kingdom. The Geological Survey of Ireland was a constituent part under the first Local Director, Captain Henry James. The Irish survey was first located in the Customs House, Dublin. Its designated task was to produce a map of the geology of Ireland that improved on the Griffith map of 1839. De La Beche issued 'Instructions for Local Directors of the Geological Survey of Great Britain and Ireland', including directions that the mapping was to be conducted on a scale of 1:10,560 (6 inches to the mile), using as a base the topographic maps of the Ordnance Survey and starting in the southeast of the country. These base maps would then be adjusted to final versions for the public at a scale of 1:63,360 (1 inch to the mile). Southeast Ireland was chosen as a starting point for two reasons. First, the Survey already had a team mapping directly across the Irish Sea in Wales, and so there could be collaboration over common problems. Second, since a strong reason for geological mapping in Ireland was the search for economic minerals, mapping the known occurrences of coal at Castlecomer in Co. Kilkenny and copper in Avoca, Co. Wicklow would emphasize the economic benefits of the new Survey.

Further instructions laid out the method to be used for mapping in the field. The geologist would record in pencil the rock types at each exposure and make notes and sketches in a field notebook. This information would be inked in that evening and the collected information duplicated on a second set of sheets. Periodically the geologist would also draw the essentials onto a set of two one-inch scale maps which would be sent to Dublin for approval on completion. Following approval, the geologist placed all the information on a clean copy of the one-inch map for publication, and the completed map was sent to the Ordnance Survey for engraving. Following any amendments, the geologist added the standard colours for each rock type to make the pattern copy, which was the standard to which copies were coloured for sale to the public. This task was usually undertaken by the Survey colourists in England (see Figure 3.11).

One such person was Mrs E. Williams, whose job it was to turn the uncoloured sheets into what can

Figure 3.11 One-inch, Sheet 32 (Pettigoe), 1885, Survey colourist, Mrs E. Williams, Camden Town, London (Extract from horizontal or longitudinal section 25. Section from Kilkieran Bay on the south to Killary Harbour on the north, 1879. Geological Survey of Ireland, 1:63,360 series; Geological Survey of Ireland, geological map 1:63,360 series). Courtesy of GSNI.

arguably be regarded as a work of art. Gordon Herries Davies, in his book, *Sheets of Many Colours*, dealing with the history of the mapping of Ireland's rocks, tells Mrs Williams's story:

> *Mrs E. Williams of 4 Stratford Villas, Camden Square, Camden Town, London.* During the closing decades of the nineteenth century she was employed to work at home colouring those maps of the Geological Survey of Great Britain and Ireland which were destined for distribution to the public. As we contemplate one of the early copies of that colourful final Irish one-inch sheet – Sheet 10 – it is Mrs Williams's handiwork that excites our admiration. Toiling in the grimy air of Victorian London, close to the Metropolitan Cattle Market and hard by the Midland Railway's line out of St Pancras, hers was a landscape far removed from the Irish landscape of purple mountains, tranquil rivers, sandy eskers, brown peat-bogs, and storm-beaten cliffs – an Irish landscape which had become so familiar to generations of geologists as they painstakingly traced out those geological lines which Mrs Williams's brush now had to follow with such care and precision. (p. 229)

Hand colouring continued until around 1900, after which the sheets were colour printed.

The production of a map such as Figure 3.12 was the mainstay of Survey activity until nearly the end of the twentieth century, by which time, following Partition, the geological Survey of Ireland had been joined by the Geological Survey of Northern Ireland, based in Belfast, and staffed by geologists from the British Geological Survey. The traditional mapping methods

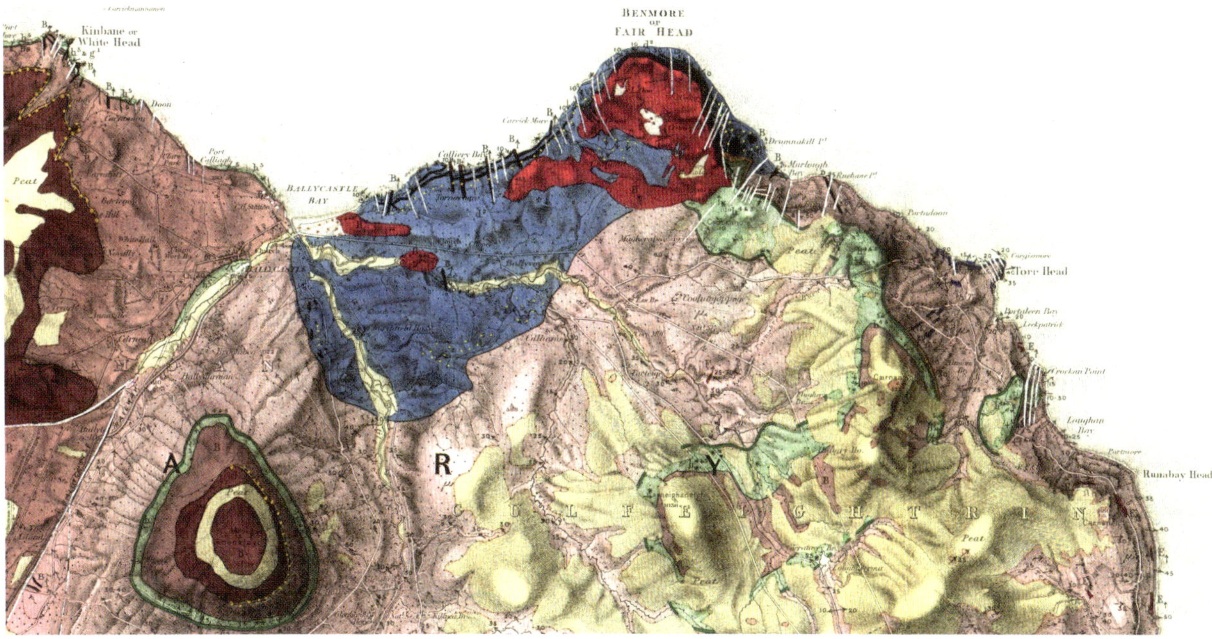

Figure 3.12 Extract from Geological Survey of Ireland, Sheet 8, Ballycastle, August 1887. Courtesy of GSNI.

employed by survey officers in the nineteenth century were still in use, supplemented by innovations involving geophysics and geochemistry, and opportunities for aerial surveying. By the first decade of the twenty-first century, the scene was set for a new phase of representation of the landscape of Ireland. This time feet on the ground collecting soil and water samples would be supplemented by eyes in the sky, a survey aeroplane carrying a whole range of detectors to remotely analyse characteristics such as the chemistry, geomagnetism, resistivity and conductivity and radiation levels – the Tellus Survey.

The Tellus Survey

In 2004, the Tellus Survey was officially launched in Belfast by the Geological Survey of Northern Ireland as the first phase of a project intended eventually to cover the whole of the island of Ireland. The goals were environmental, educational and commercial as well as purely scientific. These included advancing Ireland's natural resource industry by facilitating new mineral discoveries; providing a baseline of information against which to measure environmental change; helping comply with modern legislation requirements on soil, water and resource monitoring; detecting and mapping forms of industrial and agricultural contamination and understanding the conditions in which these might develop. This survey could be said to cover the aims of the original geological survey that started in the nineteenth century, while being expanded to cover the priorities of a modern society concerned about the welfare of the environment around us. The scope of the work included a detailed airborne *geophysical* survey covering the whole land and inland water areas of the island; a *geochemical* soils survey and a geochemical *stream-sediment* and *stream-waters* survey.

The observations that provide the basis of the various surveys that have been discussed so far in this account have all been in the visible light spectrum and made via the eyes of the various survey officers carrying out their duties. It is of course possible to view the landscape in other ways than through human eyes, and the Tellus Survey has done just that. The Tellus aircraft used wing-borne detectors capable of measuring changes in the magnetic field and electrical conductivity of the Earth's surface, while other detectors on board measured levels of natural radiation.

Figure 3.13 Ternary image of three radiometric components: potassium (K), thorium (Th) and uranium (U) (values high in all three components appear as bright white, and values depleted in all three appear dark. A predominance of red, blue or green indicates a proportionate increase in potassium, uranium or thorium, respectively). Courtesy of GSNI.

These airborne detectors allow us to see the landscape in a different way, just as an X-ray image allows us to view the interior of a body and presents a different image compared with the view in visible light. This measures the landscape in terms of its natural *gamma radiation*, which is a penetrating form of electromagnetic radiation arising from the spontaneous decay of atoms such as uranium, potassium or thorium. This decay process is referred to as *radioactivity*. All three components may be amalgamated into a single ternary image (see Figure 3.13), which colour-mixes the three components, such that signals high in all three appear as bright whites and signals deplete in all three appear dark. A predominance of red, blue or green indicates a proportionate increase in potassium, uranium or thorium respectively.

The source of most terrestrial gamma radiation is radioisotopes of potassium, uranium and thorium. Radioisotopes are atoms with unstable nuclei, which undergo radioactive decay, giving out energy, including gamma rays. Almost all gamma radiation is absorbed by 35 cm thickness of rock or soil, so the observed signals characterize only this thin surface layer. Despite this limitation, the data can be used in differentiating rock types, in the prospecting of radioactive minerals and in assessing amounts of in-house radon, a gas derived from uranium and thorium, exposure to which is linked to health problems, including lung cancer. The ternary distribution, above, of the three radiometric components, presents a coloured image of the landscape viewed through the radiation detectors carried in the survey plane. The very dark areas in the northeast largely correlate with the outcrop of the Antrim Lava group, which are low in radiogenic elements, while the bright areas in south Co. Down coincide with the granites of the Mourne Mountains, high in radiogenic elements. Areas such as south Down and south Armagh have a higher potential for Radon gas accumulation because of their higher levels of radiogenic elements.

Geophysical Survey

Geophysics is the study of the physical properties of the Earth. The airborne geophysical survey is capable of measuring three of those properties from a moving aircraft. These are:

- the magnetic field;
- the electrical conductivity;
- radioactivity.

These remote sensing techniques are especially useful in regions such as Ireland where the bedrock is obscured by extensive areas of superficial deposits such as glacial till, alluvium or peat.

The Magnetic Field

Most rocks are slightly magnetic and differences in the measured magnetic field indicate variations in the type of rock and soil beneath the aircraft. These

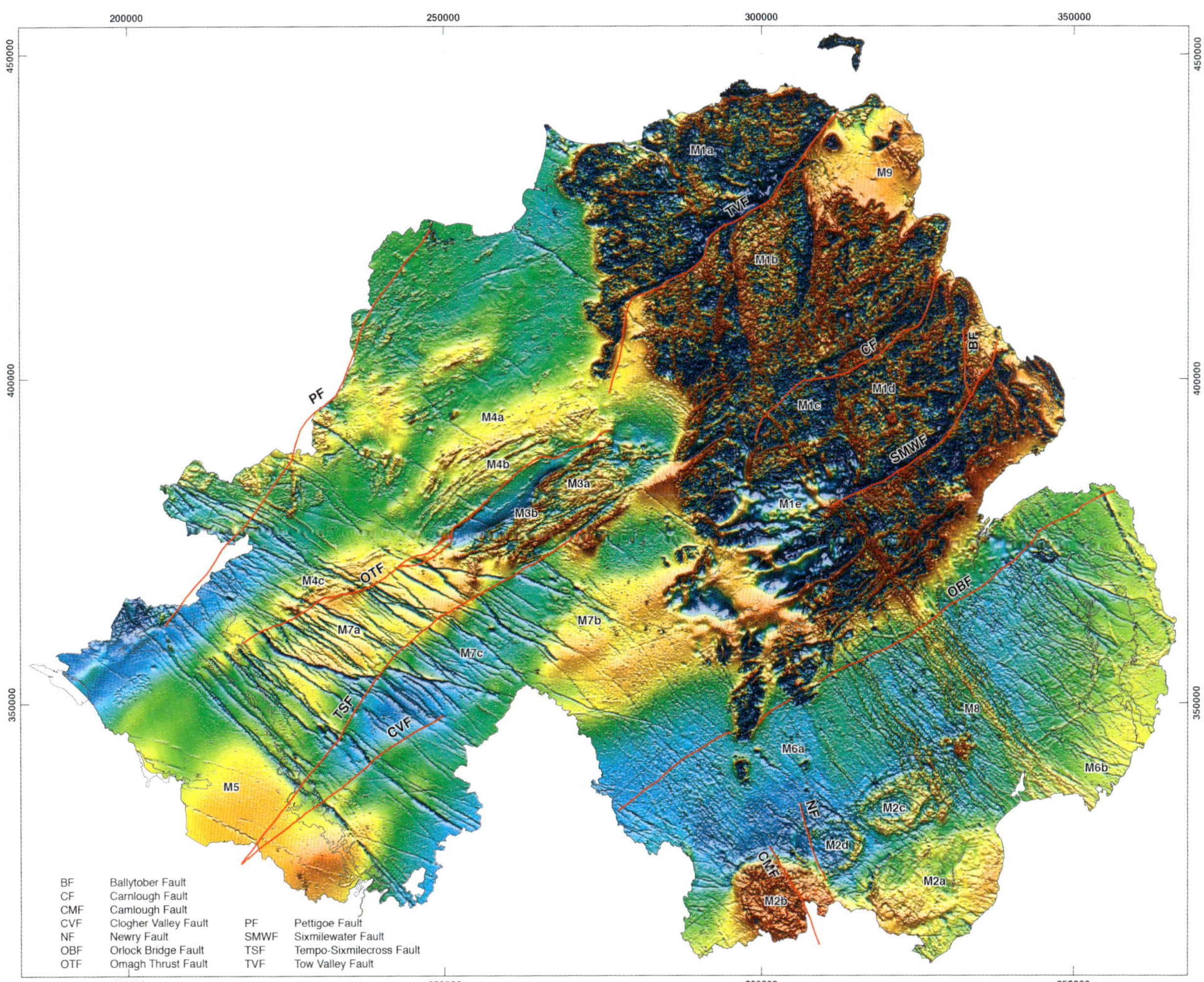

Figure 3.14 Total magnetic intensity: blues and greens are low values (nT), yellows and reds are high values. Courtesy of GSNI.

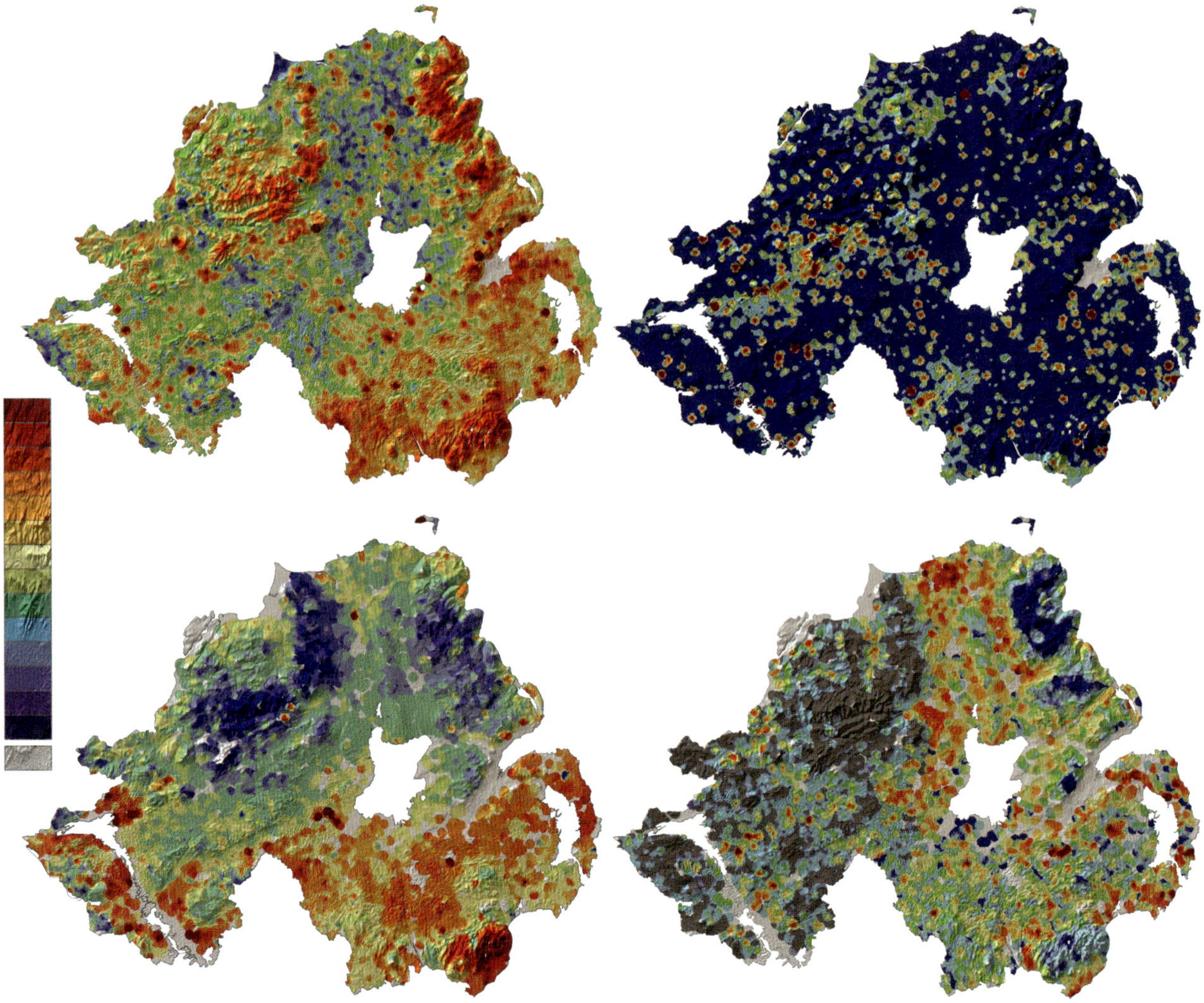

Figure 3.15 Examples of element distribution maps, Tellus Survey, Northern Ireland. Top left, lead in top soil; top right, gold in subsoil; bottom left, fluoride in stream water; bottom right, nitrate in stream water. Courtesy of GSNI.

variations indicate not only the relatively shallow effects of magnetic rocks near the surface, but also major features of the geology deep within the crust. Major features of the magnetic field over Northern Ireland include the Antrim Lava Group, a swarm of liner intrusions, dykes, in Fermanagh, and the granite intrusions of the Mourne Mountains in south Co. Down (see Figure 3.14). Prominent regional faults such as the Tow Valley Fault in north Antrim are also clearly mapped.

Geochemical Survey

Geochemical data from the rocks and soils and from stream samples can serve several different functions. They can define a standard or baseline for current levels of an element, they act as an aid in prospecting for economically important minerals and they detect anomalous levels of potentially harmful elements and compounds in the soil or in stream waters. For example, the images in Figure 3.15 show lead in topsoil, gold in subsoil, fluoride in stream water and nitrate also in stream water.

- **Lead** in topsoil: lead is biologically non-essential and well known for its toxic properties. It can cause mental impairment in young children and neuropathy and hypertension in adults.

- **Gold** in subsoil: with the high demand and consequent high cost of the metal, gold mineralization is always an active field for exploration and research and the Tellus geochemical surveys have shown the widespread distribution of anomalous concentrations of gold, both in deep soils and in stream sediments throughout the region. This suggests a local bedrock origin and the Tellus data are being used in efforts to find further economic gold deposits.
- **Fluoride** in stream water: Fluorine is an essential micronutrient for mammals, where it strengthens skeletal tissues and teeth. Low levels may be detrimental to dental health, but high concentrations may induce skeletal disorders.

These are just some of the uses of the wide range of available geochemical results that have already come from the Tellus Survey. Data collection is ongoing, and 75 per cent of the island was scheduled to have been surveyed by 2020.

4

Chronicles of the Landscape

There is a long tradition of accounts of the Irish landscape by 'naturalists' – people who described the 'natural history' of the country – the flora, the fauna, the rivers and lakes, the shoreline, the topography and the distribution of the various types of bedrock.

One of the earliest of these writers, and perhaps one of the most atypical, was Giraldus Cambrensis, or Gerald of Wales, (*c.*1146–*c.*1233), a Welsh-Norman cleric and retainer at the court of king Henry II of England. In 1185, he was chosen to accompany one of the king's sons, Prince John, on John's first expedition to Ireland. Gerald's work, *Topographia Hibernica* (also known as *Topographia Hiberniae*), first published in 1188, is his chronicle of the landscape and people encountered on his journey around Ireland.

This was his second visit to Ireland, the first being a few years earlier, in 1183. His family were important in the Norman administration of Ireland for the English Crown, and he was acting as an advisor to his uncle and brother. This first stay was certainly not more than a year's duration and in 1185 he landed in Ireland for the second time at Waterford, in the southeast of the island. During this first visit there is no evidence he went beyond the districts of Waterford and Cork and on his second trip he travelled from Waterford to Dublin, probably by boat or at least by a coastal route. Other locations we can be reasonably sure he visited were Arklow, Wicklow, Meath, Kildare and perhaps the River Shannon near Athlone and Lough Derg, also on the Shannon, near Limerick.

The book falls into three parts, the first of which has most relevance to this study, dealing with the Position of Ireland. The second and third parts deal with the *Wonders and Miracles of Ireland* and the *Inhabitants of the Country*, respectively. The Position of Ireland covers the distance between Ireland and Britain, the lands to the north and south and the latitude and longitude of Ireland. The topography of the country is commented on, and the distribution of the principal rivers and lakes noted. There are comments on the flora and fauna of the countryside. Ireland is depicted as a fertile land, but the population was barbarous and semi-pagan in nature. Many of the descriptions are anecdotal and record events and fauna which to modern eyes seem to be fantastical or whimsical. To some commentators, this detracts from the credibility of his account, but its depiction of the native population as somehow debased allowed the invaders to assume a particular moral position with respect to them that was used to justify the conquest and subsequent occupation of Ireland over the next several centuries. It is also worth considering that *Topographica Hibernica* is one of very few available sources describing life in twelfth-century Ireland.

Arthur Young (1741–1820) was an English writer on agriculture, economics and social statistics, and established a reputation as an expert on agricultural reform. He travelled extensively in Britain, Ireland and France, and published his *Tour in Ireland: With General Observations on the Present State of that Kingdom* in 1780, after spending the years from 1776 to 1779 touring the country. This was republished in 1892 as *Arthur Young's Tour in Ireland (1776–1779)*, edited by Arthur Hutton.

Young made comprehensive observations on many aspects of life in Ireland towards the end of the eighteenth century. These included comment not only on the farming methods employed in various parts of the country but also remarks on the condition of the roads and other public works, the variation in rents and wages and a section dealing with manners and customs of the population. It is true to say he is not always impressed by what he sees in comparison with conditions in England. Much of the tone of his commentary can be summed up by the image of an Irish *Cabbin* [*sic*], a thatched Irish cabin with smoke issuing from the door and vegetation spouting from its roof. The dwelling has no windows or chimney, light enters via the door and the smoke from the fire makes its way out how it will. Young says the cottages of the Irish 'are the most miserable looking hovels that can be well conceived', consisting generally of one room, and mostly bereft of substantial items of furniture. The book contains much comment on the politics and the social, economic and religious condition of Ireland at the time, and, like the work of Gerald of Wales, is a unique and valuable source of information on life in Ireland at the time.

Following the publication of Arthur Young's travel account in 1780, a series of similar travelogues describing the Irish landscape were published in the final years of the eighteenth century and into the nineteenth century. One example, which followed in 1786, was *Letters Concerning the Northern Coast of the County of Antrim*, by Reverend William Hamilton (1755–97), a Protestant minister who had a keen interest in geology, meteorology and archaeology. His book included observations of the various geological features of the North Antrim area, including the Giant's Causeway and the Carrick-a-Rede volcano. His comments on the basalts of the Giant's Causeway contributed to the eventual recognition of the role of volcanic activity in their origin as advocated by the *Vulcanists* in their dispute with the *Neptunists*. In brief, *Vulcanists* believed that rocks such as at the Giant's Causeway originated at high temperature from deep in the Earth, while the *Neptunists* believed that most rocks crystallized from seawater. See Chapter 6 for a more detailed discussion of the *Vulcanist*/*Neptunist* controversy.

The locality known as Hamilton's Seat, on the clifftop east of the Giant's Causeway, is reputedly where he rested on his walks exploring the geology of the area. It is notable that he included an itinerary for visiting the Giant's Causeway from Belfast, a prescient thought indeed, given the numbers who nowadays flock there all year round.

Hamilton came to an unfortunate and premature end. His parish was in north Donegal, where he was also a magistrate, and thus seen as a representative of the British government. Tensions between tenants and landowners that would eventually be an important factor in the 1798 Rebellion caused his death at the hands of an angry mob who attacked the house he was staying at near Manorcunningham in Co. Donegal.

The second half of the nineteenth century saw the advent of photography, and with it the opportunity for a more graphic account of the landscape, and one which could be used to provide a permanent record. In addition, society saw a growing interest in and awareness of the natural environment. From the 1860s onwards, there was a growth of organizations that aimed to encourage and develop this interest. In 1863, the Belfast Naturalists' Field Club (BNFC) was formed to cater for this burgeoning desire among the intelligentsia of the day in Victorian Ireland to explore the natural world. Such exploration was facilitated by the growth of the railway network and the development of the concept of leisure time and tourism. The formation of the Belfast club was followed by its Dublin equivalent in 1886, with further groups forming in Cork and Limerick before the end of the century. Figure 4.1 shows the group at the Giant's Causeway in all their Victorian self-importance. The dominant figure to the right of the picture is William Grey, an Inspector under the Office of Works, described by Robert Lloyd Praeger (1865–1953) as a 'science jack of all trades'. Perhaps a further measure of the social standing of this group is the presence of T. K. King, the figure in the hat to the

Figure 4.1 Belfast Naturalists' Field Club fieldtrip, Giant's Causeway, Co. Antrim, 1868. Photograph courtesy of the Belfast Naturalists' Field Club.

left of William Grey, listed in the BNFC archives as 'U.S. Consul'.

A major consequence of the founding of the BNFC was the early education in natural history of Robert Lloyd Praeger, who was destined to become one of the most influential natural historians in Ireland in the late nineteenth and the first half of the twentieth centuries, and who readily acknowledged the importance of his early training in geology on the BNFC field trips with the likes of Inspector Grey.

Robert Lloyd Praeger: The Way That He Went

Robert Lloyd Praeger was born in Holywood, Co. Down, the son of Willem Praeger, who had come from The Hague in Holland to Belfast to work in the linen industry. His mother was Maria Patterson, the daughter of a Belfast businessman, with connections to the linen industry. Robert showed an early interest in the natural world, and although he took a degree in civil engineering and had a position with the Belfast City and District Water Commissioners, he was more interested in the fauna he found in the estuarine clays exposed during the building of the Victoria Dock in Belfast. Over the next ten years, he made significant contributions to the geology and palaeontology of the clay, sand and peat deposits of the Belfast area, before joining the staff of the National Library in Dublin in 1893, where he remained for the rest of his career, retiring as chief librarian in 1923 (Figure 4.2).

Praeger made a massive contribution to botanical studies in Ireland, as demonstrated by awards he received, such as the Gold Medal of the Royal Horticultural Society of Ireland and honorary life membership of the Botanical Society of the British Isles. A major result of his botanical endeavours appeared in 1901, with the publication of *Irish Topographical Botany*, a landmark work in the history of Irish botany. For the first time, a detailed geographical analysis of Ireland's vascular flora was compiled. Praeger was largely instrumental in proposing and organizing the Clare Island Survey in the early years of the century under the auspices of the Royal Irish Academy. This was a multidisciplinary survey of Clare Island, off the west coast of Ireland, during 1909–11, and involving over a hundred scientists, including zoologists, botanists, archaeologists and geologists. While the bulk of the work concerned the flora and fauna, other topics were included, such as geology, archaeology, climate, place names and agricultural practices. This was a pioneering study; nowhere had such a detailed study of the natural history or cultural heritage of an area of comparable size been made. The work provided an early example of a baseline survey. This is a survey used to monitor changes in the environment revealed by subsequent surveys. Around 100 years later, the Royal Irish Academy repeated the exercise to enable an evaluation of changes on the island in the intervening period.

Figure 4.2 *Robert Lloyd Praeger*, by Sarah Cecilia Harrison. National Museums, Northern Ireland. Art UK. https://commons.wikimedia.org/wiki/File:Robert_Lloyd_Praeger_by_Sarah_Cecilia_Harrison.jpg.

In 1895, Robert Lloyd Praeger began a secular pilgrimage across Ireland that was to take up every weekend of the spring and summer months of his life for the next five years. On Friday evenings, he made his way from his place of work, the National Library, to Heuston Station (then Kingsbridge) to board the Great Southern and Western Railway or Midland Great

Western Railway services, which at that time connected to nearly 500 rural railway stations across the country. Saturday and Sunday were spent botanizing with a purpose. His aim was to chart the distribution of all plants across the country – nothing less than a one-man census of the entire Irish flora. By Sunday evening, he had walked across country or conducted a 40-mile circuit that brought him to a railway station and a return train to Dublin for work on Monday morning. He could do this on 'a tooth-brush and a collar' and the plants he had collected would still be fresh when he got home.

New Survey of Clare Island, 1991–1995

A resurvey of Clare Island has been undertaken by the Royal Irish Academy. A total of 26 fieldwork studies have been completed, encompassing five disciplines: archaeology, history and culture, botany, geology, and zoology. The survey is a major Academy venture, bringing together more than 100 experts from Europe and all over Ireland to document changes in aspects of the island's heritage, environment and biology.

The photograph of Croaghmore on Clare Island (Figure 4.3) allows the introduction at this stage of the narrative of another significant figure in the recording of Irish natural history during the late Victorian era. The development of photography from the mid-nineteenth century onwards was a significant factor in the recording of the Irish landscape. Among the pioneers of this new technique was Robert John Welch (1859–1936). Born in Strabane, Co. Tyrone, he established his photography business in Belfast in 1883 and started a career that specialized in outdoor photography and encompassed contemporary life as seen in the rural and urban landscape of Ireland. He covered a wide range of topics and illustrated such publications as railway company guidebooks as well as scientific journal articles on subjects as varied as geology, archaeology and natural history. In his role as official photographer for the shipbuilders Harland and Wolff, in Belfast, he produced an iconic series of images chronicling the building of the RMS Titanic and her sister ship, RMS Olympic (Figure 4.4).

Welch was the official photographer for the Clare Island Survey. A near contemporary with Praeger, they were both members of the Belfast Naturalists' Field Club and must surely have overlapped at lectures or fieldtrips in Belfast before Praeger moved to Dublin in 1893. Welch's photographs were among the first to be used to illustrate articles in learned journals. For example, the Knockmany Passage Grave, a burial monument in Co. Tyrone, was photographed by Welch to illustrate a paper by George Coffey published in 1898, and in 1897 the unconformity between the Triassic rocks and the Cretaceous chalk at Murlough Bay in North Antrim was photographed by Welch to illustrate a paper by W. Frazer Hume, published in the *Quarterly Journal of the Geological Society of London*, the leading geological journal of the time. A very early example of the use of photography in academic publications, the classic geological textbook, *Principles of Physical Geology*, by Arthur Holmes, first published in 1944, contains a number of Welch's photographs.

The early collaboration between Welch and Praeger allowed Welch to become involved in what many would regard as Praeger's magnum opus – his natural history travelogue around Ireland, *The Way That I Went*, published in 1937. The book included the image of Croaghmore (Figure 4.3), taken by Welch during the Clare Island Survey.

Following his retirement from the National Library in Dublin in 1923, Praeger continued travelling and exploring around Ireland, amassing information on the flora, fauna and geology of the whole country. Along with the development of photography, the spread of the railway network greatly facilitated a project of this nature.

Praeger's book is a fascinating example of the role played by essentially amateur naturalists in the Victorian era and into the early twentieth century. The original Clare Island Survey, completed in 1911, was inspired by Praeger in his role as a natural historian. The survey and *The Way That I Went* cover a wide range of disciplines, including botany, zoology, geology and archaeology, yet Praeger was a civil engineer by original training.

His book was a resounding success, with the most recent edition (1997) brought out to commemorate the 60th anniversary of its original publication. Its greatest impact, however, was the influence over the next generation who came to write about the Irish landscape in all its aspects – people like E. Estyn Evans, Frank Mitchell and Tim Robinson.

(1951), *Irish Folk Ways* (1957) and *The Personality of Ireland: Habitat, Heritage and History* (1973), which together offered a unique treatment of the links between anthropology, geography, history and the influence of the landscape. It is interesting that he dedicated his *Mourne Country* to Robert Lloyd Praeger and acknowledged his example and inspiration. Like Praeger, Estyn Evans believed in fieldwork as a way of developing an intimate understanding of the landscape and the way of life of its inhabitants. *Irish Folk Ways* represents a record of past practices and customs, which were fast disappearing in an Ireland that was modernizing at an escalating rate.

As well as leaving a substantial literary legacy, Estyn Evans was also the driving force behind the creation of the Ulster Folk and Transport Museum at Cultra in Co. Down. He was the founding director of the Institute of Irish Studies at Queen's University, Belfast, which promotes and co-ordinates research in fields of study with a particular Irish interest. His last major publication, *The Personality of Ireland: Habitat, Heritage and History*, dealt with the interaction of these three strands, which allowed a holistic interpretation of the Irish landscape and its regional character and sense of place. As discussed in Chapter 1, by habitat he meant the total physical environment, by history he meant the written record of the past, and heritage he defined as the unwritten segment of human history, man's physical, mental, social and cultural inheritance from a prehistoric past. This definition included oral tradition, beliefs, languages, art and crafts. The book serves as a comprehensive view of Ireland and is a fitting summary of his work over the previous four decades.

As a first-year undergraduate at Queen's in the mid-1960s, I had the privilege of being taught by Estyn Evans. In a classic example of youth being wasted on the young, I did not appreciate my good fortune, and to this day I regret not taking full advantage of being in the presence of such an erudite scholar whose insights on the history of landscape development in Ireland still resonate almost 100 years after he first disseminated them. However, what I did take from my encounter with Professor Estyn Evans was that an all-encompassing or holistic approach to the interpretation of landscape development was essential to its understanding.

Frank Mitchell and *The Way That I Followed*

A near contemporary of Estyn Evans, who was also heavily influenced by Praeger, was Frank Mitchell (1912–97). Like Praeger, Mitchell came from a comfortable background, in his case on the south side of Dublin, brought up in an environment where education was important and an awareness of nature part of family life. He was keenly interested in birds and was a member of the Dublin Naturalists' Field Club, where he first encountered Praeger, who was a regular participant at their meetings. After graduating in Natural Sciences from Trinity College Dublin (TCD), his first position was as assistant to Professor Knud Jessen of Copenhagen, who was leading a project investigating the history of the development of bogs in Ireland and the formation of the glacial sediments that they covered. This was taking a pioneering multidisciplinary approach to the problem and paved the way for his lifetime interest in the integration of the various fields such as pollen studies, geomorphology, geology, archaeology and the evolution of human settlement in Ireland. This led eventually to his appointment to the personal Chair of Quaternary Studies at TCD in 1965. The Quaternary Period in geological history is from approximately 2.6 million years ago to present and includes the Pleistocene, which covers the most recent Ice Age in Ireland.

Mitchell's book, *The Way That I Followed: A Naturalist's Journey Around Ireland*, published in 1990, was a conscious nod to Praeger's *The Way That I Went*, which had come out more than fifty years earlier. Mitchell's rationale was that not only had the country changed radically in the interim, but that many scientific techniques were now available which were previously unknown, and he was interested in examining the contrasts in the scientific environment between his time and Praeger's. In contrast to the route followed by Praeger, who 'zig-zagged down Ireland', Mitchell followed what he described as a 'continuous if winding route'.

The Way That I Followed is a very personal account of the natural history of Ireland and contains many autobiographical aspects. It can probably be thought of as an act of homage to Praeger's innovative study, both of them based on a philosophy whereby the best way of understanding the country was via the soles of one's boots. This work followed Mitchell's pioneering account of the Irish landscape, *The Shell Guide to Reading the Irish Landscape*, published in 1986, which was then revised, in collaboration with Michael Ryan, an archaeologist, and republished in 1997 as *Reading the Irish Landscape*. This revised and updated work is a tour de force, which deals with the making of the modern Irish landscape using a multidisciplinary approach to cover Ireland's geology, geography, archaeology and history. It is a more formal treatment than the *Shell Guide* and deals chronologically with the history of the Irish landscape from its formation to the present. Paradigm changes in the Earth Sciences such as Plate Tectonics allow Ireland's geological history to be placed in a global context. The development of techniques such as *palynology* (pollen studies) and *dendrochronology* (the study of tree-rings) assist in understanding the evolution of the landscape from the eventual retreat of the ice and the arrival of the earliest settlers. The changes from the Neolithic through the early Christian period to the Norman invasion and the Plantation are documented. The chronological sequence records the evolution of Irish society following the Famine and the impact of the Industrial and Agricultural Revolutions of the nineteenth century. The immense changes in Ireland in the latter half of the twentieth century, particularly flowing from its membership of the European Union, are discussed in terms of the impact on the Irish landscape. In all, the book is a remarkable compilation and contains in a single volume the story of the Irish landscape in a seminal treatment of a whole range of related disciplines.

Geography often serves as an umbrella discipline, encompassing a range of fields essential to understanding complex phenomena. In tracing the evolution of Ireland's landscape – from its geological origins to its post-glacial occupation by *Homo sapiens* around 10,000 years BP – we have been fortunate to benefit from the work of two distinguished geographers, Emyr Estyn Evans and Frank Mitchell, who contributed significantly and contemporaneously to this area of study. Both were strongly influenced by Robert Lloyd Praeger, who was by inclination a naturalist in the Victorian amateur tradition. Praeger imbued those who worked with him with the idea that boundaries between traditional specialisms should be fluid and that only by collaboration and teamwork could the complete picture be realized.

Tim Robinson and *The Stones of Aran*

Following the pioneering work by Praeger describing the characteristics of the Irish landscape, a labour of love that was carried on and extended by Estyn Evans and Mitchell, the next to detail aspects of our environment was Tim Robinson (1935–2020). Robinson was an Englishman, brought up in Yorkshire, where as a child he explored the countryside around Wharfedale and Ilkley Moor. A Cambridge University graduate in mathematics and physics, he had a career as a visual artist in Istanbul and Vienna before returning to the UK in the 1960s and working as a technical editor while continuing to exhibit his artwork. In the 1970s, he and his wife moved to the west of Ireland where he began a detailed study of the landscape of Connemara, the Burren and the Aran Islands over the following 40 years or so (Figure 4.5). Over this period, he wrote several innovative and influential books, which were accompanied by detailed hand-drawn maps. In the two volumes of *The Stones of Aran*, the first, *Pilgrimage*, traces the perimeter of Árainn, the largest of the Aran Islands, also known as Inishmore, describing the flora, fauna and physical features of the sea and rocks, and commenting on the cultural history of the area. This exploration of the cultural history places a great deal of emphasis on the oral tradition and the Irish language, and relies heavily on those residents, often elderly, who accompanied him on his walks. The second book, *Labyrinth*, deals with the interior of the island in the same way. His later studies of Connemara continue this theme of the relationship between the environment and the inhabitants of the area.

The geology and landscape of the Aran Islands is a continuation of that found in the Burren region of Co. Clare on the nearby Irish mainland – that is, a karstic limestone topography, weathered into near-horizontal terraces, separated by softer layers of mudstones and shales. The islands were separated from the mainland at the end of the Ice Age about 10,000 years ago when the sea level rose as the ice melted.

On the relevant section of his map, Robinson points out the bay called Port Bhéal an Dúin, which means

Figure 4.5 View of central Inishmore from Dún Aengus, Co. Galway. Ingo Mehling. https://commons.wikimedia.org/wiki/File:Inishmore_Central.jpg.

'the port of the fort's mouth', the fort being Dún Aengus (Dún Aonghasa), three-quarters of a mile to the west (see Figure 4.6). The spectacular Dún Aengus is the best known of the hill forts on the Aran Islands and lies at the edge of a 100 metre-high cliff, whose defences consist of three walls, a band of limestone pillars known as *chevaux-de-frise*, and the sea to the south. It is not known exactly when Dún Aengus was built, probably it dates from the Bronze or Iron Ages.

Throughout his travels on the islands and in Connemara, Robinson was acutely aware of the centuries of conflict over land ownership, culture and language. As discussed in the section dealing with the tensions between the local population and the officers of the Ordnance Survey in the late eighteenth and early nineteenth centuries, as they anglicized the place names for their maps, the sense of place felt by long-standing inhabitants of a region is a powerful cultural influence.

By recording the lore surrounding the physical features of the landscape in his books, and by adding many of these details to his maps, he hoped to preserve this aspect of the cultural heritage in a world that was changing all too quickly. It is clear from Robinson's work that an enormous store of information concerning the geology and hydrogeology, the history, the culture and folklore of an area resides within the place names in the original Irish. He has performed an epic service by recognizing and recording this information, much of it via the oral tradition before it was lost irrevocably as older residents died and the knowledge died with them.

Figure 4.6 Dún Aengus Fort, Árainn, Co. Galway. Ronan Mac Giollapharaic. https://commons.wikimedia.org/wiki/File:Dun_Aonghasa_(cropped).JPG.

Tim Robinson's endeavours to preserve the community knowledge of the landscape of Árainn bears resemblance to the concept of 'songlines', as performed by First Nation or Indigenous Australians. Their 'creation myths' tell of the legendary totemic beings who wandered over the continent in the 'dreamtime', singing out the name of everything that crossed their path. In this way, the birds, animals, plants, rocks and waterholes of the landscape were sung into existence. The oral history of places is carried in song cycles and a person with the relevant knowledge is able to navigate across the land by intoning the words of the song that describes the location of the various landscape features and other natural phenomena. While some of these songlines are a few kilometres long, others cover hundreds of kilometres and were used by Indigenous people to navigate across the Australian continent.

Robert Flaherty and *Man of Aran*

In looking at the work of Robert Flaherty on Aran, the same aim of preserving the cultural heritage in a rapidly changing world is apparent but is carried out using a very different and sometimes controversial medium. Robert Flaherty was an American documentary filmmaker who first came to prominence in 1922 with the release of his silent documentary film, *Nanook of the North*, which followed an Eskimo (Inuit) family for almost two years on the eastern shore of Hudson Bay in eastern Canada as they travelled in search of food, hunting walrus and seals, building an igloo and visiting the white man's trading post. Modern society is now so used to television documentaries and travelogues about and from remote regions of the world that it can be difficult for us to fully appreciate the impact this film had at the time of its release. That this represented groundbreaking cinema is not in doubt, and it was well received by critics and became a box office success in the USA and elsewhere. It shone a light on the culture of the indigenous people living in a harsh environment in a location far removed from so-called civilization. Controversy arose as it was realized that in places in the film Flaherty had staged some of the events, for example that while the man who played Nanook normally used a rifle while hunting, the walrus-hunting sequence showed a harpoon, and that in the seminal igloo-building sequence the interior shots were in fact shot in a special three-walled igloo to allow enough light for filming. In his defence to these charges, Flaherty asserted that it is often necessary to distort something in order to reproduce its true spirit. It is perhaps ironic that the film was backed financially by a French fur-trading company, whose activities were among those hastening the demise of the very way of life being portrayed in the film.

Flaherty's endeavours in Arctic Canada became relevant to the Irish landscape with his next major project. *Man of Aran* (1934) was a silent documentary film about living on the Aran Islands, written, directed and shot by Flaherty. It shows what purports to be the daily activities of a family on the islands, farming, fishing and collecting seaweed. The cinematography is spectacular and shows off the remarkable landscape and seascape of the island and the surrounding Atlantic. As with *Nanook of the North*, *Man of Aran* generated much controversy over the degree to which events were staged. For example, a scene was contrived in which the fishermen are almost lost at sea in a sudden gale, and the boat is destroyed by the waves on the shore. Nor is the family genuine: they were residents chosen for their photogenicity.

When it premiered in Dublin, in 1934, the film was well received by the government of the Irish Free State. It was seen to reflect the image and values of the country emerging as an independent country after centuries of rule from Britain – a rural, traditional, resilient culture. In contrast, its critics accused it of including fabrications and distortions and presenting an image that ignored the real problems of the country – particularly in the west and on the islands – the absentee landlordism, the isolation, the poverty and the ever-present curse of emigration.

Just as the traditional Inuit way of life was vanishing in northern Canada, much of the lifestyle portrayed in *Man of Aran* had already been overtaken by modern developments. Yet the film effectively highlights a landscape unique within Ireland and demonstrates how the physical environment shapes cultural responses. Since Flaherty's time, public opinion has become far less tolerant of anything resembling staging, re-enactment or manipulation – practices now widely regarded as ethically problematic.

5

Art in the Landscape

The perception of the landscape depends not only on the physical attributes imposed by the bedrock characteristics and geomorphic processes, but it also depends on who is living the experience. Those such as artists, sculptors, writers and architects (I use the term architect very loosely here) who are viewing the landscape from an aesthetic point of view will have a different set of priorities from those who derive their living directly from that same landscape as farmers, stonemasons or miners. Changes in the thought processes of *Homo sapiens* that occurred about 70 kya, known as the 'cognitive revolution', led our ancestors to develop the ability to show creativity and imagination. What followed these changes was the appearance of figurative art, music, dance and rituals around cooking and burial in what is known as the Palaeolithic Period, the Old Stone Age, from around 2.6 mya to 12 kya. These changes represented the beginnings of civilization.

In the Blombos Cave, South Africa, are what are claimed to be the oldest known drawings by human hands, estimated to be around 73 ky old. In 2002, the recovery of two finely engraved ochre pieces was reported. The surface of both pieces had been intentionally modified by scraping and grinding to form an engraved pattern of cross-hatching in combination with parallel incised lines. See Figure 5.1.

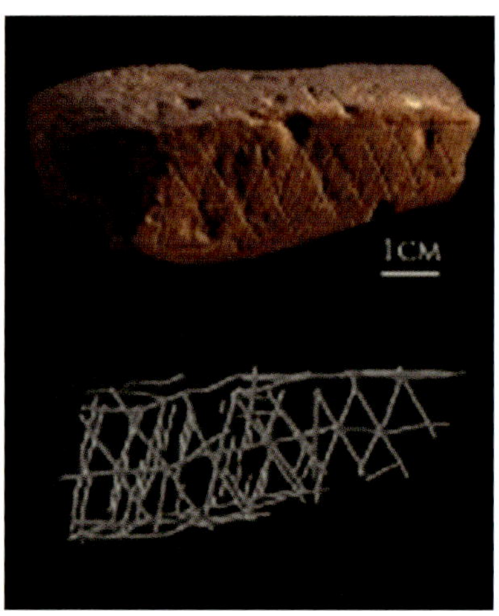

Figure 5.1 Possibly the oldest known drawing by human hands, Blombos Cave, South Africa (estimated to be 73,000 years old). Chris S. Henshilwood. https://commons.wikimedia.org/wiki/File:Blombo.jpg.

Figure 5.2 Chauvet Caves, Ardèche, southeastern France. HTO, Public domain, via Wikimedia Commons. HTO. https://commons.wikimedia.org/wiki/File:Lions_painting,_Chauvet_Cave_(museum_replica).jpg.

Among the best-known and best preserved early art is the Chauvet-Pont-d'Arc Cave in the Ardèche department of southern France, a cave that contains some of the best-preserved figurative cave paintings in the world, as well as other evidence of Upper Palaeolithic life. The paintings are dated to between 30000 BCE and 28000 BCE (Figure 5.2). The change from the earliest societies of the Palaeolithic to those of the Neolithic (New Stone Age) was caused by a warming world, as the effects of the Ice Age receded and the practice of agriculture spread. Figure 5.3 presents a timeline of early settlement in Ireland.

The 'Stone Age' is divided into the Palaeo-lithic, the Meso-lithic and Neo-lithic periods – the Old, Middle and New Stone Ages. The Palaeolithic is considered

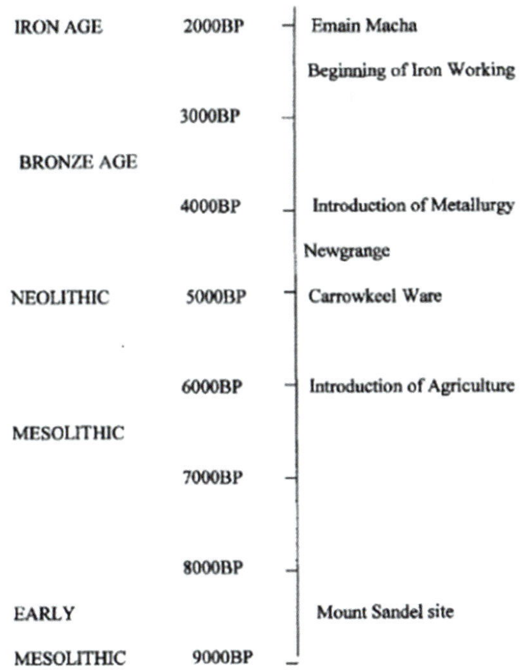

Figure 5.3 Timeline for settlement in Ireland, before present.

to cover the period from 2.6 mya to 10 kya, succeeded by the Mesolithic, lasting from 10 kya to 6 kya. The Neolithic succeeded the Mesolithic around 6 kya until the Bronze Age, which began around 4 kya.

The earliest known settlers of Ireland appear to have arrived sometime around 10 kya, in the Mesolithic (10 kya to 6 kya), probably by sea from Britain. They were hunter–gatherers who lived off seasonal nuts and berries, hunting the available fauna and fishing the coastal waters and inland rivers and lakes. One of the best known Mesolithic sites, mentioned earlier, is at Mount Sandel in Co. Londonderry, and shows the remains of timber shelters, hearths for warmth and cooking, and evidence of a diet comprising fish, including eels and salmon, hazelnuts, berries and meat from wild boar. This phase of settlement is also characterized by flint blades in the form of 'microliths'. These were commonly mounted in rows on wooden handles or shafts and used for cutting and sawing, and as spear heads for hunting. The ready availability of flint nodules from the Cretaceous chalk outcrop in northeast Ireland must have been a factor in the settlement pattern of the island.

The first farmers arrived around 6000 kya in the Neolithic, and with them came new ideas about food production and the ability to raise crops and domesticate animals such as cows and sheep. It appears extensive forest clearance took place at this stage, opening up space for farming and leaving signs of large field systems such as the Céide Fields in Co. Mayo. During this period, megalithic tombs were introduced for burial rituals. Examples such as the passage tombs occurring in so-called 'cemetery' clusters, such as at Carrowkeel and Carrowmore in Co. Sligo and Newgrange and Loughcrew in Co. Meath, are extremely impressive, with quite complex rock art in the form of spirals and chevrons. See discussion below.

Rock Art

Rock art is the name applied by archaeologists to a form of prehistoric art prevalent in northern and Atlantic Europe. In Britain and Ireland in the Neolithic, many carved rocks are found featuring cup and ring motifs. The term 'petroglyph' (from the Greek *petros*, rock, and *glyphein*, to carve) is often used to describe these features. The image is created on the rock surface by a process of incising, picking, carving or abrading the rock, thereby removing material. In Ireland it is best seen in certain coastal localities in Kerry and Donegal and south Leinster, and in western parts of Ulster. One of the most impressive examples is found at Reyfad, Boho, County Fermanagh. The Reyfad Stones consist of six stones, five of which bear curvilinear cup-and-ring markings, dating from the Neolithic to the Iron Age – approximately 5,000 years ago (Figure 5.4). Comparable megalithic art appears at Newgrange,

Figure 5.4 Reyfad prehistoric art, Boho, Co. Fermanagh. Courtesy of NIEA.

Knowth, and Dowth in County Meath, which dates to around 5,200 years ago.

Monuments in the Landscape

In Ireland, societal changes were coming into effect by about 6 kya, with the transition from a mainly hunter–gatherer mode of life to a more sedentary life around agriculture. Among these changes was the interment of the dead in cemeteries, which sometimes featured *megalithic* tombs. It was in this context that rock art and the culture of megalithic monument building developed. As mentioned in the Introduction, the building of megaliths represents a recognition of *time*. The tombs are built in the *present* and represent a measure of communal effort and co-operation. They are monuments to the *past* in the form of ancestral remembrance, and they look to the *future* in that they were clearly intended for future generations and to outlast those responsible for their construction.

According to Gabriel Cooney, in his *Landscapes of Neolithic Ireland* (2000), *monumentality* is the physical embodiment of a social order and may have been intended primarily to create a sense of permanence in the relationship between communities and the landscape. While there may be a link between the spread of agriculture and a corresponding increase in the time available for the communal effort required to build such monuments, there appears to be some overlap with the earlier hunter–gatherer lifestyle. Richard Bradley, in *The Significance of Monuments* (1993), suggests that monuments located in places already considered special may be designed to emphasize this specialness rather than intrude on such spaces. The building of

Figure 5.5A Court cairn. Rob Hurson, CC BY-SA 4.0 https://creativecommons.org/licenses/by-sa/4.0, via Wikimedia Commons.

Figure 5.5B Portal tomb or dolmen. Courtesy of NIEA.

Figure 5.5C Passage tomb. Shane Finan, CC BY-SA 4.0 https://creativecommons.org/licenses/by-sa/4.0, via Wikimedia Commons.

Figure 5.5D Wedge tomb. Slongy, CC BY 4.0 https://creativecommons.org/licenses/by/4.0, via Wikimedia Commons.

monuments, which involved the selection, movement and erection of large masses of rock, is an indication that the people of the Neolithic had considerable understanding of the mechanics and mathematics of the engineering required.

It is possible to recognize four major types of megalithic tombs in the Irish landscape (Figures 5.5A–D).

- *Court cairns* are shaped in a long mound, with a forecourt at one end leading into a long chamber (e.g., Creevykeel, Co. Sligo).
- *Portal tombs*, formerly known as *dolmens*, consist of a number of upright stones covered by one or more capstones (e.g., The Giant's Ring at Ballynahatty, near Belfast).
- *Passage tombs* are round mounds, having a burial chamber roughly in the centre, reached by a passage leading from the edge of the mound (e.g., the Carrowkeel complex in Co. Sligo).
- *Wedge tombs* have long rectangular chambers, usually roofed with large stones and placed in a long wedge-shaped mound (e.g., the Altar edge Tomb near Schull in Co. Cork).

Figure 5.6 Newgrange passage tomb decoration. Johnbod, CC BY-SA 3.0 https://creativecommons.org/licenses/by-sa/3.0, via Wikimedia Commons.

Cooney (2000: 16) writes that a notable feature of the passage tomb tradition is the occurrence of megalithic art, with rectilinear and curvilinear motifs, particularly in Bru na Boinne. There are affinities between some of the circular motifs of megalithic art and those typical of rock art discussed earlier in the chapter. This suggests that rock art motifs, which are inscribed on rock outcrop surfaces, began in the Neolithic and continued into the Early Bronze Age, when it is thought the decorated standing stones at Newgrange were added.

Passage grave art includes carved motifs that differ from those seen in exposed examples such as at Boho. These include single and double spirals, diamond- or lozenge-shaped patterns, triangles and zig-zag lines (see Figure 5.6).

Visibility in the Landscape

There are obvious differences in the degree to which monuments are prominent in the landscape. For example, not all are placed on high ground: some clearly are designed to blend into the landscape, while others

Figure 5.7A Concordant passage tomb, Carrowkeel, Co. Sligo.

Figure 5.7B Kilclooney More portal tomb, Co. Donegal. Andreas F. Borchert, CC BY-SA 3.0 DE https://creativecommons.org/licenses/by-sa/3.0/de/deed.en, via Wikimedia Commons.

are intended to dominate an area. This difference has been described as '*sympathetic*' and '*constructive*' respectively (Cooney, 2000). The term *concordant* could be applied to describe monuments in harmony with the adjacent landscape, while those set up to transform the landscape are *discordant*.

Passage tombs, such as at Carrowkeel in Co. Sligo (Figure 5.7A), are considered to be in harmony with the landscape and the adjacent rock outcrops that are the source of the rocks making up the tomb. In this case, the concordant monuments are used to emphasize or enhance some feature of the landscape already considered to be important. This contrasts with monuments such as the Kilclooney More portal tomb in Co. Donegal (Figure 5.7.B), or the Poulnabrone dolmen (Figure 1.5), discussed in the Introduction, which are clearly designed to produce a striking and discordant effect on the landscape. At Poulnabrone, the monument is constructed on a flat limestone pavement using locally derived slabs of limestone, but even so the effect is one of a radically changed landscape. In contrast, some monuments use rock types from outside the immediate area to produce an imposing new landscape. In the complex of monuments at Carrowmore in Co. Sligo, the uprights of the *Listoghil* megalith are composed almost exclusively of metamorphic rocks, probably originally from the nearby Ox Mountains, while the capstone is limestone and the whole edifice sits on limestone bedrock (Figure 5.8).

The Sligo-based artist Patricia Curran Mulligan played an important role in the archaeological investigation of the Carrowmore complex, including producing illustrations of the various artefacts recovered from the tombs. In addition, she has produced evocative paintings of the landscape around this part of Co. Sligo. See Figures 5.9A and 5.9B for her interpretation of the landscape around Carrowmore.

Figure 5.8 Listoghil megalith, Carrowmore, Co. Sligo. Limestone cap-rock resting on metamorphic uprights.

Figure 5.9A Carrowmore and Knocknarea, Co. Sligo, by Patricia Curran Mulligan. Courtesy of the artist.

Figure 5.9B Carrowmore Cemetery, Co. Sligo, with Knocknarea in the background.

The geology of the countryside around Sligo is dominated by flat-lying sequences of Carboniferous limestones and shales, characterized by the spectacular form of Ben Bulben (Figure 5.10). The area around Ben Bulben is often referred to as 'Yeats Country', as it features prominently in the poetry of W. B. Yeats (1865–1939). Yeats was an Irish poet, dramatist and writer, awarded the Nobel Prize in Literature in 1923. He was also a prominent political activist who served two terms as a Senator in the government of the Irish Free State. In his poem, *The Celtic Twilight*, he refers to this part of Co. Sligo: 'But for Ben Bulben and Knocknarea,/Many a poor sailor'd be cast away'.

Figure 5.10 Ben Bulben, Co. Sligo. Jon Sullivan (pdphoto.org). https://commons.wikimedia.org/wiki/File:Benbulbenmount.jpg.

His poem *Under Ben Bulben* is a description of the landscape of Yeats Country

Under bare Ben Bulben's head
In Drumcliff churchyard Yeats is laid.
An ancestor was rector there
Long years ago, a church stands near,
By the road an ancient cross.
No marble, no conventional phrase;
On limestone quarried near the spot
By his command these words are cut:

Cast a cold eye
On life, on death.

Horseman, pass by!

Yeats is buried at nearby Drumcliffe Church at the foot of Ben Bulben, and the last three lines of the poem are inscribed as his epitaph.

Standing Stones and Stone Circles

With the advent of the Bronze Age, megalithic tombs were no longer being built, rather the masonry skills of the Neolithic people were preserved by the erection of standing stones and stone circles. Along with these were stone pairs, stone rows or alignments. These newer monuments were composed of stones that were set apart rather than together and were not associated with cairns. The standing stone is a widely used form of monument in many different cultures. Clearly *anthropomorphic* interpretations are possible with monoliths, as in the *inuksuit* monuments constructed by the peoples of the circumpolar regions, particularly northern Canada and Greenland. These are piles of stones stacked in the rough form of a human figure. In the areas around the high Arctic regions, they have

Figure 5.11 Inuksuit monuments, Baffin Island, Canada. Ansgar Walk, CC BY-SA 2.5 https://creativecommons.org/licenses/by-sa/2.5, via Wikimedia Commons.

traditionally been used as landmarks and as indicators of special places in the landscape (Figure 5.11).

The Beltany Stone Circle in Co. Donegal is composed of more than 60 uprights, although there may originally have been many more. Outside the ring there is a single free-standing stone, a monolith, which is about 2.5 metres high. Unlike the fine-grained metamorphic rocks comprising the stone circle, the standing stone is an igneous rock, coarse-grained, containing a white mineral and a greenish mineral in roughly equal proportions. When freshly quarried, it would have been very different in appearance when compared with the circle uprights, although these differences have been obscured by thousands of years of weathering and a covering of lichens. One possible source for the monolith is a small intrusion on Croaghan Hill some 5 kilometres away.

The Beltany Stone Circle is set within a ring of low hills up to 6 kilometres away, so that the circle is a small-scale representation of the surrounding landscape (see Figure 5.12). In this it resembles stone circles such as the Ring of Brodgar in the Orkney Islands, where the circle mirrors the landscape and the theme of circularity in the landscape is common.

The Beaghmore Stone Circles in Co. Tyrone is one of the most extensive stone complexes in Ireland. It comprises seven circles, a series of stone alignments and several small stone cairns (Figure 5.13). A remarkable feature of this complex is that until the late 1930s it lay hidden under a peat bog and was only revealed as the peat was gradually removed for fuel. It is a spectacular example of how the landscape is constantly evolving. When the people of the Neolithic and Early Bronze Age, around 4,000 years ago, were erecting these monuments, the area was free of peat bog. The climatic changes discussed earlier, which led to changes in vegetation and clearance of woodland by early farmers, led to the development of *blanket bogs* in the higher ground and *raised bogs* in lower areas.

The Neolithic can be regarded as typified by the building of megaliths such as the various types of tombs found in the landscape, often associated with cairns, while the Bronze Age was characterized by the erection of standing stones and stone circles. In the succeeding Iron Age, (500 BCE–400 CE), much of the artwork was in metal, as seen in the La Tène style of decoration. This style is thought to have been introduced to Ireland from Europe, where iron had become the principal metal

Figure 5.12 Beltany Stone Circle, Co. Donegal. Mark McGaughey, CC BY-SA 4.0 https://creativecommons.org/licenses/by-sa/4.0, via Wikimedia Commons.

Figure 5.13 Beaghmore Stone Circles, Co. Tyrone. Wax0nightmare, CC BY-SA 4.0 https://creativecommons.org/licenses/by-sa/4.0, via Wikimedia Commons.

used for tools and weapons by the Celts. La Tène is an abstract curvilinear style of decoration and named after a site in northwest Switzerland where the artifacts showing this style were first found

While much of the artwork was in metal, the tradition of working in stone continued. One of the best examples is the Castlestrange Stone in Co. Roscommon, dating from the Iron Age, probably to the last few centuries BCE. The ornament is similar to that found on La Tène metal work.

Figure 5.14 Castlestrange Stone, Co. Roscommon. https://commons.wikimedia.org/wiki/File:IMG_CastlestrangeStone2771.jpg. Sarah777 at en.wikipedia, Public domain, via Wikimedia Commons.

The Coming of Christianity

Following the Iron Age, the next phase of Rock Art in Ireland was associated with the establishment of Christianity in the fifth century. One of the most evocative features of the Irish landscape is the high cross, representative of the early medieval church. A superb example is Muiredach's Cross at Monasterboice, Co. Louth (Figure 5.15).

Figure 5.15 Muiredach's Cross, Monasterboice, Co. Louth. Adriao, CC BY-SA 4.0 https://creativecommons.org/licenses/by-sa/4.0, via Wikimedia Commons.

When the first free-standing ringed crosses appeared in the eighth century they heralded a completely new phase of Irish art in the landscape. It is possible to plot the development of these elaborate monuments via a series of earlier types. Pillars inscribed with crosses can be seen in places such as Glencolumbcille in Co. Donegal (Figure 5.16). However, unlike the later classic crosses, these early examples did not change the outline shape of the original rock. They are often composed of schist, with relatively soft mineral mica as a major component. This softness means it can be carved with relatively shallow cuts.

Figure 5.16 Decorated standing stone, erected in the Christian era, Glencolumbkille, Co. Donegal.

Figure 5.17 Earliest free-standing cross at Carndonagh, Co. Donegal.

Many of the earlier standing stones in the area were cross-inscribed during the Christian era. The village of Glencolumbkille and its surrounding area is today well known for the *Turas*, a pilgrimage around a number of *stations* consisting of cross pillars, cairns, megalithic tombs and natural landscape features, such as wells and rock outcrops. The earliest free-standing cross appears to be at Carndonagh in northwest Donegal (Figure 5.17).

The cross is dated to the seventh century since the pattern on the cross is similar to the *Book of Durrow*, an illuminated manuscript of around 650 CE that consists of text from the four Gospels. The Carndonagh Cross is important to the understanding of the evolution of high crosses in Ireland. This early example of a Christian cross is thought to represent the transition from crosses carved on slabs to a slab that is carved in the shape of a cross.

The later high crosses normally consist of three elements, *a base*, the *shaft* and a *capstone*. The shaft usually incorporates the *ring*. The cross at Ardboe, in Co. Tyrone (Figure 5.18), was built in the tenth century, and shows many of the features typically found as the design evolved into a more sophisticated form.

Figure 5.18 High Cross at Ardboe, Co. Tyrone. Henry Clark, Ardboe Cross. https://commons.wikimedia.org/wiki/File:Ardboe_Cross.jpg.

After an early stage of high cross development showing abstract decoration, scenes from the Bible were carved on the faces. Later on, high crosses were commonly erected around monastic sites and with their biblical scenes they were probably important sites in the religious calendar of the community centred on the monastery. The crosses may have been used as teaching aids for religious instruction in a pre-literate era. The crosses are usually sculpted from sandstone, but examples exist carved from limestone and granite.

The ancient Egyptians developed techniques to produce large blocks of rock from quarries in order

Figure 5.19 Unfinished obelisk at Aswan, Egypt. Diego Delso, CC BY-SA 4.0 https://creativecommons.org/licenses/by-sa/4.0, via Wikimedia Commons.

to manufacture tall monuments known as obelisks. These monuments originated around 1400 BCE and were prominent in their architecture, playing a vital role in religious ceremonies. An unfinished obelisk found at Aswan (Figure 5.19) gives some insight into the techniques used by the Egyptians to quarry such massive blocks of stone from the bedrock.

The unfinished obelisk at Aswan is the largest known ancient obelisk, around 42 metres long and weighing over 1,000 tonnes. As work proceeded on carving it out of the bedrock, cracks appeared, and the project was abandoned. Similarly, in Ireland, the Unfinished Cross at Kells (Figure 5.20) shows that the crosses were probably removed in large roughed-out blocks from the quarries and that the final carving happened after the block was put in place.

Unlike the earlier inscribed stones, such as at Kilnasaggart, the later high crosses do not seem to have had a role as a funerary memorial. Their function is clearly different and their association with monasteries was probably a function of the right of sanctuary afforded by monastic sites. They also had a role as assembly points for religious ceremonies and the later versions with their carved biblical scenes may have served as an aid in the teaching of scripture, either to a lay audience or to members of the monastic community.

Figure 5.20 The Unfinished Cross at Kells, Co. Meath. Stephen Keaveny, CC BY-SA 4.0 https://creativecommons.org/licenses/by-sa/4.0, via Wikimedia Commons.

What Purpose is Served by the Erection of Stone Monuments in the Irish Landscape?

It is interesting that the erection of high crosses in Ireland was part of the growth and development of Christianity but evolved out of earlier perceptions of sacred or special features in the landscape. These features were singled out for endorsement or modification by the community, perhaps in an attempt at establish some sense of permanence in their affinity with the landscape. Like the utilization of Neolithic sacred sites by the later Celts in the Iron Age, so those in the early Christian era took over sites in the landscape considered special by earlier cultures. The desire to modify the landscape with pillars or buildings of one form or another to commemorate the past or the present has continued from the Neolithic to the present day.

As recently as 1999, a granite monolith, the Strangford Stone, quarried from the Mourne Mountains, was erected in Co. Down to commemorate the new millennium. A single block of granite was quarried from the flanks of Slieve Donard, and like the unfinished obelisk at Aswan (Figure 5.19), the quarried block split into two parts during the extraction process. Unlike at Aswan, however, the project was not abandoned. Instead, the two parts were rejoined using stainless steel dowels.

This monolith was manually hauled into position at Delamont Country Park and raised upright (Figure 5.21). Its south face bears an engraving depicting the exact alignment of the four inner rocky planets – Mercury, Venus, Earth and Mars – on the day the stone was erected. The project drew upon traditional skills and communal effort, echoing the practices used in monument building since prehistoric times. These included quarrying techniques for extracting large

Figure 5.21 Strangford Stone (granite monolith), Co. Down. Ardfern, CC BY-SA 3.0 https://creativecommons.org/licenses/by-sa/3.0, via Wikimedia Commons.

stone blocks from bedrock, shaping the monolith to the desired dimensions, and carving it with appropriate decorative motifs. Finally, transporting the stone from its point of origin to the chosen site required sophisticated engineering methods and the coordinated labour of over a thousand individuals

The standing stone represents the simplest of monuments and as such has a widespread distribution across many cultures and over long time periods. With the advent of written language, such as in Christian-era Ireland, it is possible to be certain of the motives for the erection of a specific monument. This serves as an aid to the interpretation of earlier monuments, erected before the development of written language. There seems, however, to be a common thread running through these monuments from earliest Neolithic, when they were first erected, to the present day. A desire to place on the surface of the Earth, often in a prominent position in the landscape (but not necessarily so) some form of upright marker. This marker designates that place as special, or it may commemorate a person, a group or an event. That the culture of standing stones, menhirs, obelisks, totems and all the other variations is so widespread geographically and has endured from the earliest cultures to the present day is surely an indication of the common roots of civilization that we all share.

6

Artists and the Landscape

This book is an attempt to integrate the physical features of the Irish landscape with other responses from a range of perspectives, which can loosely be described as 'cultural'. These responses can be by individuals, be they artists, sculptors, writers or poets. Or they can be community responses, as in the formation of rock art, or the building of megaliths. Part of that cultural response can be in the form of using the landscape as a resource for the betterment of society as a whole. An example of that would be the Neolithic use of certain rock types to produce stone axes. These axes were then used to clear the forests more efficiently and allow the expansion of agriculture. There can be a focus on prospecting for a resource in the form of a suitable material to produce stone axes, and there can be at the same time the aesthetic appreciation of landscape as a source of psychological well-being.

Geologists see the landscape differently from non-geologists. This is largely because of their perception of time. Unlike most other scientists, geologists must factor the passage of time into their deliberations. For them, any view of the landscape is translated into the sequence of events that formed the landscape. Through the long history of the Earth this requires an understanding of the age of the component rocks and landforms making up the scenery in question. Geologists read the landscape in their efforts to tell this story of formation and development. An inherent problem is that much of the evidence is missing and so geologists are always solving problems with incomplete data. As part of this process, geologists often talk about the picture or sometimes the big picture, meaning the sum of all the components involved. These include factors such as the movement of tectonic plates, the types of rock present, the fossil assemblages and climatic influences. The picture put forward will be severely constrained by the available evidence and any such model may have to be modified in the event of additional facts coming to light.

A geologist reading the Irish landscape is attuned to both long-term and short-term changes that continue to shape it. Although tectonic plates remain in motion, the relative absence of significant earthquake activity here can lead us to overlook these ongoing processes in everyday life. These are transformations that have unfolded over billions of years and will persist far into the future. One possible consequence for the Irish landscape is the eventual closure of the Atlantic Ocean and a renewed formation of mountain ranges akin to the Caledonian Mountains, which arose during the closure of the Iapetus Ocean (see Chapter 2).

Superimposed on these vast geological timescales are more recent and visible changes. The retreat of ice following the last glaciation has left an unmistakable imprint, yet even these glacial features have been altered by successive waves of human settlement since the arrival of Mesolithic peoples. From the shift to agriculture and the transformative effects of the medieval agricultural revolution, to the industrialization of the mid-nineteenth century, each phase of human activity has left an indelible mark on the Irish landscape.

This progression has been recorded in various ways by artists using a variety of methods and media. The rate of change in modern times is such that a new geological epoch was proposed. At the time of writing, the proposal to formally codify a new epoch, the *Anthropocene*, has been rejected by the International Commission on Stratigraphy, the relevant scientific body which adjudicates on such matters. However, the term has entered the public domain and there is a widespread recognition that humans now represent the

Figure 6.1 *Below Ground*, by Edward Ferguson. Courtesy of the artist.

principal geological agent on the planet. Modern artists thus respond to a landscape that is changing in ways that are different to those viewed by their predecessors in the eighteenth and nineteenth centuries.

In contrast to a geologist, an artist contemplating painting a picture of that same landscape has a much greater degree of freedom as to how the view is interpreted and need not be concerned about any subsequent changes. This is seen in Figure 6.1, where the Co. Down artist Edward Ferguson has imagined an underworld based on geological features visible at the surface and extrapolated to depth.

In addition, a variety of motivations may have influenced an artist in the process of making a representation of the landscape. If the painting has been commissioned in earlier centuries by a wealthy patron, many landscape artists will have sought to please that patron with a flattering vision of the house or estate. By contrast, others may have had an agenda involving a nationalist view and so may have wished to portray an authentic *Celtic* landscape by referring to traditional practices and customs of rural Ireland. The political turmoil in Ireland during the nineteenth century and into the twentieth century produced a range of images representing the formation of the new Irish state, the societal changes involved in the Industrial Revolution and the development of a modern industrial society. This search for a national identity has been a major motivational force not in only Irish art but also in Irish literature. For example, authors such as W. B. Yeats and J. M. Synge wrote about aspects of the way of life that existed in the more remote, western parts of the island.

The varied geology of Ireland underpins a wide range of scenery across the island. A number of these landscapes can be regarded as totemic and are seen by outsiders as typically Irish. Examples of these

Figure 6.2 Blue plaque at birthplace of Paul Henry, in Belfast. Peter Clarke at English Wikipedia.

iconic landscapes are the paintings of Paul Henry (1876–1958) of the west of Ireland (see Figure 1.1). Henry, born in Belfast (Figure 6.2), spent time in the west of Ireland, notably on Achill Island, and created a body of work that in the 1920s and 30s did much to create the popular view of the west of Ireland. As discussed in the Introduction, the use of Paul Henry landscapes by various railway companies to advertise holidays in Ireland did much to create this stereotypical view of the Irish landscape.

Other artists moved to the west, including another Belfast man, Gerard Dillon, who made his first visit in 1939. Like Paul Henry, he was inspired by the grandeur of the landscape but also fascinated by the way of life and customs. The largely unspoken truth, however, is that this west of Ireland idyll was more apparent than real. The image existed largely because of the extreme poverty caused by the neglect of the area and its people by the central government in Dublin. Life in the west of Ireland during the first half of the twentieth century was harsh, with most people relying on subsistence farming and fishing. Families were typically large, and for many young men and women, emigration was the only viable alternative. What had been a way of life born of necessity for the local population became, for visiting artists, a source of imagery that evoked a vanishing past – one that appeared to preserve a connection to traditions increasingly under pressure as Ireland evolved as a new nation.

Much has changed since the 1920s and 30s, when artists such as Paul Henry and Gerard Dillon were crafting their romantic visions of the Irish west. With Ireland's accession to the European Union and its transformation into a high-tech economy, regions like Connemara are no longer seen as poverty-stricken backwaters defined by emigration. Today, they are thriving communities, integrated into a modern Ireland that plays an active role within the EU.

The rise of mass travel following the Second World War and the subsequent growth of tourism have further reshaped the region. The landscapes once depicted by artists like Henry have become major attractions, forming the foundation of a vibrant tourism industry that now benefits the entire island.

It is possible to recognize that certain parts of the Irish landscape are regarded by both residents and visitors alike as being 'special places'. Like the earliest settlers drawing attention to specific localities by choosing them as sites for megalithic monuments, so these areas have become essential stops on any visitor itinerary. Such locations can be considered 'iconic'.

Iconic Landscapes

The Giant's Causeway

One of the most iconic landscapes in all of Ireland is the north Antrim coast, and specifically the Giant's Causeway (Figure 6.3A). The near-vertical columns, with their roughly hexagonal cross-sections, the result of the way the basalt lavas cooled, create a spectacular landform. While columns such as these are not unique (other examples occur in Iceland and the USA, for example), their occurrence here at the coast on a wave-cut platforms backed by high cliffs makes them easily accessible. This part of the north Antrim coastline is one of the most popular tourist destinations in Ireland.

Figure 6.3A Columnar basalts of the Giant's Causeway, Co. Antrim.

Figure 6.3B Lower Basalt Formation, Interbasaltic Formation (laterites) and Causeway Tholeiite Member.

The Giant's Causeway may now be one of the best known geological features in the world, but because of the relative remoteness of the north Antrim area it was not until the end of the seventeenth century that it was first referred to in print, and it was not marked on any map until 1714. The earliest known illustration is an engraving of an aerial perspective by Christopher Cole, published in the *Philosophical Transactions of the Royal Society* in 1694. The drawing gave some indication of the form of the columns but was highly stylized. A second article, in the same volume, by Thomas Molyneux, a prominent Irish physician and Fellow of the Royal Society, passed comment on the Causeway and commissioned Edwin Sandys to produce a more realistic version. An engraving of this was subsequently published in the *Philosophical Transactions* of 1697. While this representation was an improvement on the earlier version by Cole and gives a better understanding of the complex nature of the jointing, the field relationships are vague and confusing.

More than forty years were to pass before the next major advance in recording the nature of the jointing in the rocks of the Causeway area and their configuration in the field. Susanna Drury (1698–1770) was a relatively unknown painter from Dublin, who, with assistance from the Dublin Society (later the Royal Dublin Society), spent several months painting in the area around 1740. Her exquisitely detailed watercolour paintings of the Causeway were important markers, not only for Irish landscape painting, but also as illustrations in a Europe-wide scientific debate on the origin of volcanic rocks (Figure 6.4).

The accuracy with which Drury has depicted the multiple jointing of the columns is remarkable, and her paintings represent the first truly believable images of the Causeway and its complexity. These views of the Causeway served to popularize the locality, and between 1743 and 1744 engravings were made by François Vivarès, born in France, but living in London (Figure 6.5).

The prints made from engravings by Vivarès were widely distributed within the scientific community in Europe, where a bitter controversy was raging over the origin of rocks. Two principal schools of thought had

Figure 6.4 *East Prospect of the Giant's Causeway*, Susanna Drury, *c.*1740. Courtesy of the Ulster Museum.

developed: the *Neptunists* claimed that all rocks, with few exceptions, had crystallized from seawater, while the *Vulcanists* or *Plutonists* believed that rocks such as granite or basalt originated at high temperatures deep in the Earth, and were intruded at depth or erupted as lava flows at the surface. Using the prints made from Susanna Drury's paintings, a French scientist, Nicolas Desmarest (1725–1815), argued that the jointing in the Causeway rocks was similar to rocks found in the Auvergne district of France, which were clearly the result of lava flows issuing from an obvious volcanic crater. This proved to be the beginning of the end of the *Neptunist* theory, and by the early part of the nineteenth century, Neptunism had been confined to the history of science.

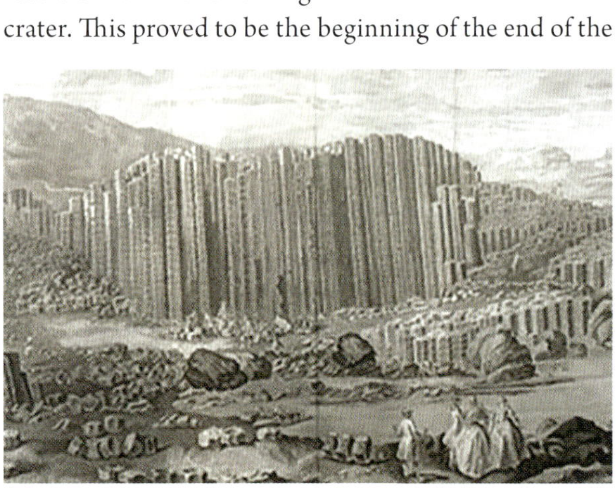

Figure 6.5 François Vivarès's engraving of the Giant's Causeway, after Susanna Drury. Susanna Drury, Public domain, via Wikimedia Commons.

Recognized as a site of major scientific interest, the Giant's Causeway attracted researchers from around the world. Simultaneously, the expansion of Ireland's railway network during the Victorian era transformed Portrush into a thriving seaside resort and made the entire north coast a popular destination for a growing and increasingly affluent population. Among the artists who contributed significantly to the visual representation of the north Antrim coast was Andrew Nicholl (1804–86). A comparison of the styles of Susanna Drury and Andrew Nicholl in portraying the same basaltic landscape of north Antrim offers a fascinating insight into differing artistic approaches (see Figure 6.6).

Compared with the meticulous attention to detail shown by Susanna Drury, the paintings by Andrew Nicholl give a much more generalized treatment. The main components of the landscape are there, the configuration of the columns and the outlines of the cliffs are discernible, but the precise nature of jointing patterns is missing. Even the figures in the landscape and the clothes they are wearing are less precise than those in the Drury paintings, and more attention is paid to the seascape, which in Drury's paintings is largely ignored.

Figure 6.6 *Giant's Causeway*, from the west, by Andrew Nicholl. Courtesy of the Ulster Museum.

The arrival of digital photography in the late twentieth century has opened a whole new chapter in landscape representation. The ability to manipulate colours and backgrounds gives much scope to produce atmospheric images, such as illustrated in Figure 6.7, by Donna Rowley.

Figure 6.7 *Where Giants Walk*, by Donna Rowley. Courtesy of Donna Rowley.

The West of Ireland

Chapter 2 described the consequences for the geological structure of Ireland of the closure of the Iapetus Ocean and the building of the great Caledonian Mountain Range stretching from the Appalachians to Scandinavia via Ireland and Scotland (see Figure 2.17).

In Ireland, the roots of these Caledonian mountains are found in the northwest. They can be seen in the Donegal Highlands and the Sperrin Mountains in the west of Ulster, and in the west in Connacht, mainly in Counties Mayo and Galway. The upland areas are formed of granites and metamorphic rocks and characterized by spectacular quartzite peaks such as Errigal

Figure 6.8A Errigal, Co. Donegal. Sleepyhead2, CC BY-SA 3.0 https://creativecommons.org/licenses/by-sa/3.0, via Wikimedia Commons.

Figure 6.8B Croagh Patrick, Co. Mayo. Bart Horeman, Croagh Patrick, the saddle on the western flanks. https://commons.wikimedia.org/wiki/File:Croagh_Patrick,_the_saddle_on_the_western_flanks_-_geograph.org.uk_-_605872.jpg.

in Donegal and Croagh Patrick in Mayo (see Figures 6.8A–B).

This northwestern part of Ireland, with its dramatic Atlantic coastline, cliffs alternating with sandy bays and long inlets, has some of the most popular scenery in the country. The whole region was heavily glaciated. Inland areas are characterized by bare rock surfaces, with numerous loughs and patches of bog. Artists have responded to this landscape in contrasting ways. Paul Henry has already been referred to as one known for his treatment of the landscape on a large scale. Many of his paintings have an emphasis on the rugged peaks against a backdrop of bright cumulus clouds, with the small cottages and attendant turf stacks reduced to a relatively minor feature in the foreground. His fellow Belfast artist, Gerard Dillon, tended to focus more on the details of life in the west. Henry and other artists who found inspiration in the west of Ireland around this time seemed progressively to idealize the region.

In contrast to this common theme of sky and mountains, Henry produced a study of a coastal feature common in glaciated areas but relatively rare in Ireland. Fjords are long, narrow inlets of the sea between high cliffs, and typically formed by marine submergence of an over-deepened glaciated valley. A fjord is formed when a glacier retreats and the sea fills the resulting deepened valley. Typical fjord coasts are found in Norway and South Island, New Zealand.

Killary Harbour in Co. Mayo is a fjord. The form and extent of its valley can be seen from the satellite image (Figure 6.9A). It is approximately 16 kilometres long and up to 45 metres deep (Figure 6.9B). It is the subject of one of Henry's most famous paintings, *Dawn, Killary Harbour*, painted in 1921 (Figure 6.10). The painting is considered by many to show Paul Henry at his best and represents the end of the period from 1910 to 1919 that he spent on Achill Island, where he developed the painting style that allowed him to capture the particular interaction between landscape and light in the west of Ireland.

A marked feature of many of Paul Henry's landscapes of the west of Ireland is the absence of people. This may appear paradoxical when considered alongside the establishment by the British Government in Ireland in 1891 of the *Congested Districts Board for Ireland*, whose remit was 'to alleviate poverty and congested living conditions in the west and parts of the north-west of Ireland'. The Board was finally dissolved in 1923 by the new government of the Irish Free State.

Another Belfast-born artist, Gerard Dillon (1916–71), also spent time in the west, and produced images that reflected this alternative, more crowded view of the area. Dillon first came to the west of Ireland in 1939 and, like Paul Henry, was inspired by the rugged scenery and the apparently simple lifestyle of its people (see Figure 6.11). In 1945, he painted *The Little Green Fields*, in which he weaves together the way of life, past and present, into the sometimes harsh landscape of Connemara (Figure 6.12). In the painting, Dillon manages to collate aspects of the region such as dry-stone walls, thatched cottages, agricultural labour and graveyards. His inclusion of a dolmen, a ruined

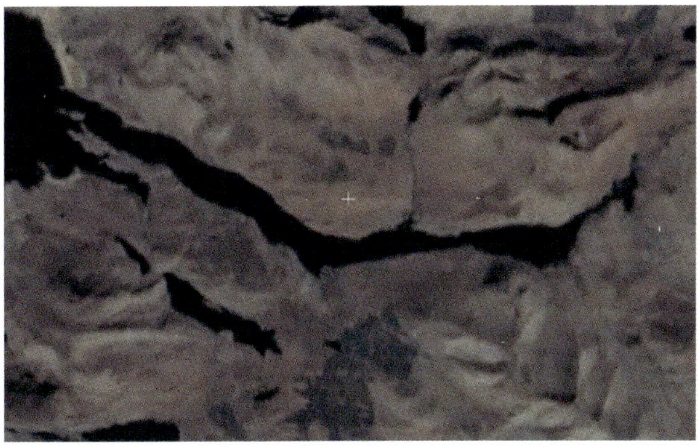

Figure 6.9A NASA satellite image of Killary Harbour, Counties Mayo and Galway. NASA, Public domain, via Wikimedia Commons.

Figure 6.9B Killary Harbour, Counties Mayo and Galway. https://commons.wikimedia.org/wiki/File:Mweelrea_%26_Ben_Lugmore_from_massif_of_Leenane_Hill.jpg.

Figure 6.10 *Dawn, Killary Harbour*, by Paul Henry (1921). Courtesy of Ulster Museum.

Figure 6.11 Bare rock outcrops, drystone walls, little fields.

of the island (Figure 6.13). The word 'Burren' comes from an Irish word 'Boíreann', meaning a rocky place. The geology is dominated by limestones of Carboniferous age, and the area is considered to be a prime example of *karst* scenery formed by the dissolution of limestones by rain and groundwater. The landscape is characterized by underground drainage systems, sinkholes and caves, and areas of bare rock known as limestone pavement (see Figures 6.14 A and B). Glaciation has stripped the area of much of the superficial deposits, such that a Cromwellian

Figure 6.12 *Little Green Fields*, by Gerard Dillon (1945). Courtesy of National Gallery of Ireland.

abbey and a high cross recognizes both the Christian and pre-Christian traditions of the area and emphasizes the super-imposition of succeeding cultures.

The Belfast-born poet Michael Longley passed comment on Dillon's paintings on the island of Inishlacken off Connemara where he spent a lot of time. Longley remarks, 'He painted the island like a seabird's nest'.[1]

The Burren, Co. Clare

If the Giant's Causeway in north Antrim and the West of Ireland can be described as iconic Irish landscapes, then so too can the Burren region of Co. Clare, in the southwest

1 Taken from 'Paintings', in Michael Longley, *The Stairwell* (London: Penguin, 2014).

Army officer (in)famously said: 'There isn't wood enough to hang a man, water enough to drown him, or earth enough to bury him'.

The Poulnabrone dolmen (Figure 1.5) is the oldest dated megalithic monument in Ireland and evidence

Figure 6.13 View of The Burren, Co. Clare. Courtesy of GSI.

of the long history of human settlement in the Burren area. The three standing portal stones support a heavy near-horizontal capstone. The monument dates from the Neolithic period, around 4200–2900 BCE.

In contrast to the cliché of Ireland as the *Emerald Isle,* characterized by the popular song, *Forty Shades of Green,* the Burren for most of the year presents a stark and barren image, as captured by artists such as Doreen Drennan (Figure 6.15).

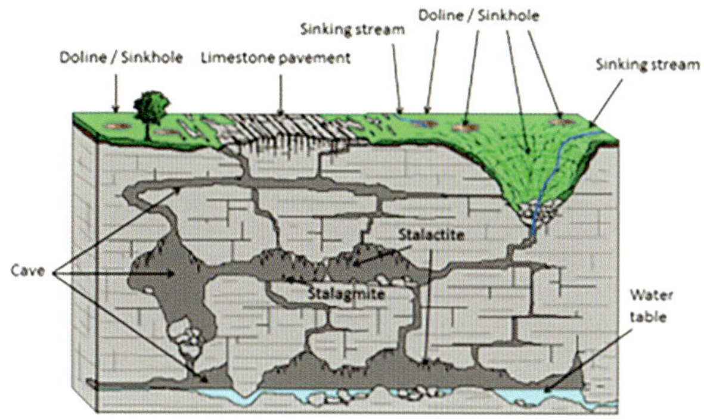

Figure 6.14A Limestone pavement, karst scenery in the Burren region, Co. Clare.

Figure 6.14B Schematic illustration of karst topography and drainage. Courtesy of GSI.

Figure 6.15 *Dawn on the Rocks.* Courtesy of Doreen Drennan.

7

Bogs in the Irish Landscape

Bogs in Ireland are a totemic feature, not only of the landscape, but also in Irish literature, painting and society in general. Bogs are defined formally as wet spongy ground, a poorly drained, usually acid area, rich in accumulated plant material, frequently surrounding a body of open water, and having a characteristic flora of sedges, heaths and sphagnum. Bogs are typically found in cool, northern climes and are associated with the aftermath of the most recent Ice Age in these areas. For the purposes of this study, we can consider two distinct types of bog – blanket bogs and raised bogs. Blanket bogs or mountain bogs are mainly found in highland areas with heavy rainfall, and they 'blanket' the entire topography with a layer of peat. Raised bogs, so called because of their elevated centres, developed in post-glacial lakes as decaying vegetation accumulated on the lakebed. This build-up eventually choked the lake until the vegetation breached the surface and eventually filled it entirely and thus formed a raised bog.

Peat is a fossil fuel that is the first stage in the long process of plant material turning into coal. Ancient bog plants, mostly *sphagnum moss*, are the fossils in peat. People have harvested peat for thousands of years. It is a source of energy for heating, insulation and electricity throughout northern Europe. Thousands of bogs throughout Ireland, Scotland, Scandinavia and Russia have been drained for peat excavation. In places like Ireland, Scotland or Scandinavia, individuals or communities have traditionally harvested peat for use as a cooking fuel. Blocks of peat are cut and left to dry in the open air before being used as fuel. In the current context of growing concern over fossil fuel use and environmental sustainability, this traditional practice is under increasing scrutiny in many communities. Nevertheless, as shown in Figure 7.1, peat cutting continues on a small scale in some areas of the island, such as here in mid-Donegal.

Figure 7.1 Small scale peat (turf) cutting, Co. Donegal.

Gustaf Hugelius and colleagues in Stockholm University analysed more than 7,000 cores drilled from peatlands in the Northern Hemisphere, measuring the

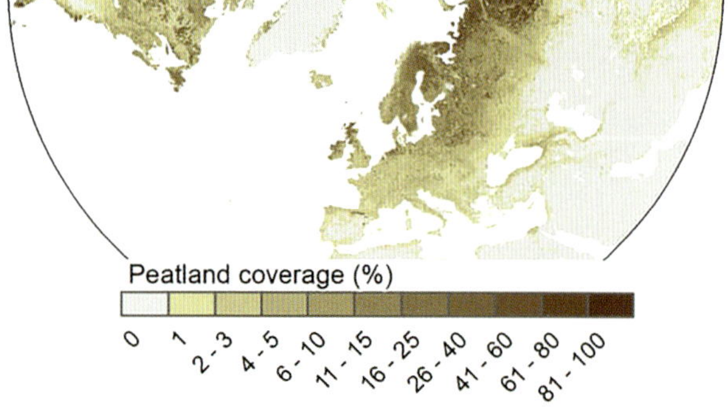

Figure 7.2 Map of the extent of peatlands in the Northern Hemisphere. Courtesy of Gustaf Hugelius, Stockholm.

depth of the peat and how much carbon and nitrogen it stored. Using these data, they created a comprehensive map of northern peatlands and showed that they covered 3,700,000 square km (1,400,000 square miles) (Figure 7.2).

The crucial importance of the northern peatlands is illustrated by their calculations that show they store approximately 415 billion tonnes of carbon – as much as is stored in all the world's trees put together. The climatic effect of the release of this amount of carbon would be devastating, and although around half of peatland carbon is locked in permafrost, global warming and its attendant permafrost thaw is likely to turn the peatland into a carbon source in the near future.

The Formation and Development of Bogs in Ireland

Bogs are among the most important habitats in the world and as such fulfil a wide range of cultural and environmental functions. Peat bogs have played a significant role in Irish culture and history. As well as providing an indigenous fuel source for the population, they preserve archaeological evidence of early human activity in Ireland, including prehistoric farming practices and artifacts such as books, weapons or boats. Bogs are also an important source of the climatic and vegetation changes in the post-glacial period as well as preserving the remains of now-extinct fauna, such as the Irish Elk or Great Irish Deer (*Megaloceros giganteus*). The distribution of the main bog types found in Ireland is shown in Figure 7.3.

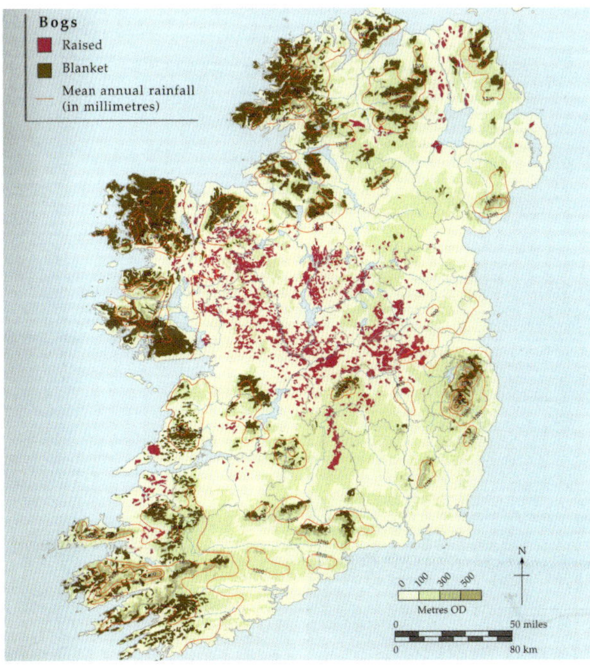

Figure 7.3 Distribution of bog types in Ireland from *Atlas of the Rural Irish Countryside.*

Formation of Blanket Bogs

Blanket bogs require a high annual rainfall and so in Ireland they typically occur in the west and in some mountainous areas (see Figure 7.4). Atlantic Blanket Bog, sometimes referred to as Lowland Blanket Bog, is found on the west coast, from sea level to an altitude of 200 metres. Mountain Blanket Bog occurs on relatively flat mountain terrains above an altitude of 200 metres. Blanket bog formation started at the end of the last glaciation, around 10,000 years ago, and is closely associated with climate change and the expansion of agriculture in the last 4,000 years. By around 6,000 years ago, Ireland had a widespread forest cover of birch and pine trees. The early settlers in the Neolithic Age began to clear these forests as agriculture expanded. The loss of forest cover meant that soils were exposed to the impact of rain, which percolated through the soil and washed, or leached, minerals such as iron downwards from the surface. This process caused the soil to become more acidic, and the

Figure 7.4 Atlantic blanket bog, Roundstone, Co. Galway. Photograph courtesy of Dr Catherine O'Connell, Irish Peatland Conservation Council.

Figure 7.5 Formation of blanket bogs. Courtesy of IPCC.ie.

iron was then redeposited at a lower level, forming an impermeable layer known as 'iron pan' or 'hard pan'. This layer prevented water from draining away, leading to the soil surface becoming increasingly waterlogged. Plants which develop in these damp conditions, such as heathers and rushes, became dominant over time, and were slow to decay because of a lack of oxygen in the waterlogged soils. This accumulation of plant debris was the beginning of peat formation and the growth of the blanket bog (Figure 7.5).

There is an increasing awareness of the importance of blanket bogs in the environment. They are not wastelands but in fact valuable wetlands with a crucial role in the regional ecosystem, including supporting a rich biodiversity of rare plants, birds and invertebrate species. Due to their capacity to store water, they act as vast reservoirs and thus play an important role in the supply of river catchments and therefore of catchment management. With increasing concern over climate change, the ability of bogs to accumulate and store millions of tonnes of carbon means they also have a role in the regulation of greenhouse gases.

Within their peat layers, blanket bogs preserve a record of their own growth and development. On a

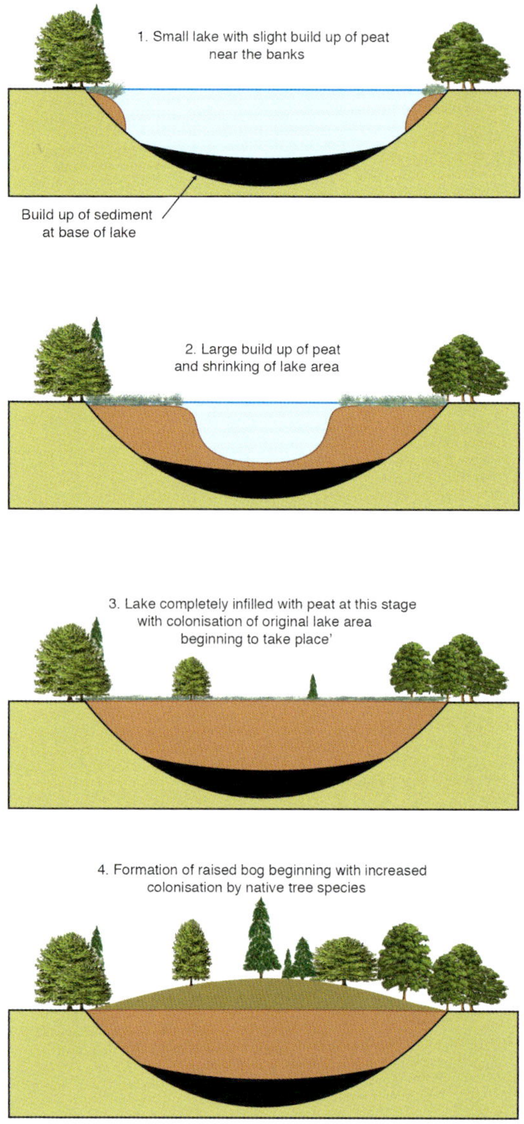

Figure 7.6 Formation of a raised bog. Courtesy of GSNI.

larger scale, they provide information on vegetation changes over time and also indicators of climate change and atmospheric pollution. The science of *palynology* is the study of pollen grains and other spores. Pollen extracted from bogs may be used for dating purposes, for studying former climate regimes and identifying plants which were then growing.

Formation of Raised Bogs

Around 10,000 years ago, as the ice retreated it left many hollows in the glacial moraines that formed small lakes. Wetland vegetation, such as reed beds, grew

Figure 7.7 Tree stumps under peat.

Figure 7.8A The Bog of Allen, Counties Kildare, Offaly, Laois, and Westmeath. Sarah777, Public domain, via Wikimedia Commons.

up around the margins of these lakes and gradually extended into the lakes. Because of the development of anaerobic conditions, complete decomposition of plant material is prevented, and the undecomposed material forms a layer of peat on the lake bed, gradually building up and choking the lake until the vegetation on the peat bed begins to emerge above the lake surface. Raised bogs are characterized by a low-growing vegetation assemblage dominated by mosses, sedges and heathers, all adapted to waterlogged and acidic conditions. Eventually, as the peat layer thickens, the roots of the plants growing on the surface are no longer in contact with the groundwater and the only source of nutrients is mineral-poor rainwater. At this stage, species such as *Sphagnum mosses* become dominant and the raised bog is established (see Figure 7.6).

For a time after the initial development of a bog, a period of drier climate meant that large trees were able to establish themselves on the bog surface. A return to wetter conditions subsequently meant that these trees were then buried by newer layers of peat (see Figure 7.7). The Bog of Allen is a large, raised bog in the midlands, covering around 1,000 square kilometres (370 square miles) across the Counties of Offaly, Meath, Kildare, Laois and Westmeath (Figure 7.8A). Until recently, peat was mechanically harvested on a large scale by Bórd na Móna, the government-owned peat production industry. As mentioned previously, peat bogs in Ireland have traditionally been used as a source of fuel. In the 1930s, the Irish government set up the Turf Development Board, which bought land under

Figure 7.8B Industrial peat harvesting in the Bog of Allen, Co. Offaly. Sarah777, Public domain, via Wikimedia Commons.

compulsory purchase orders to exploit this resource. After the Second World War, the government began to cut peat on an industrial scale for the generation of electricity, using mechanical means (Figure 7.8B). Power from peat peaked in the 1960s when it provided 40 per cent of total electricity generation. In recent years, there has been a greatly increased awareness of the importance of bogs in the environment, and there are plans to conserve the remaining areas of raised and blanket bogs.

At the time of writing (2024), Bórd na Móna has announced the biggest change of use involving Irish land in modern history. All peat harvesting is to be suspended, and all employees will be reassigned as a bog rehabilitation programme starts. The move will involve the rehabilitation of tens of thousands of acres of Irish bogs, providing new biodiverse habitats that can also support new eco-tourism and community amenity resources. The new sustainable businesses and activities located across Bord na Móna's bogs will include renewable energy development, domestic fuels, biomass development, resource recovery, horticulture, eco-tourism, and community amenities.

The Cultural Significance of Bogs in the Irish Landscape

If the hills of Connemara as depicted by Paul Henry, and the columnar basalts of the Giant's Causeway in Susanna Drury's detailed images, are considered iconic in the context of the Irish landscape, then surely the next candidate for such a status is the bog (see Figure 7.9).

To paraphrase the definition of a bog used earlier in this chapter, it is a freshwater wetland of soft, spongy ground, comprised mainly of partially decayed plant matter called peat. In less formal terms, it is an area of wet, muddy ground that is too soft to support a heavy body. The word bog is derived from *bogach* in Irish and Scottish Gaelic, meaning marsh or soft, boggy ground. It is one of relatively few words in English derived from Gaelic, and it is a measure, perhaps, of the cultural association of bogs with Ireland, that many of the derogatory terms for Irish men and women used by outsiders include the word bog – 'bogtrotter', for example.

Bogs are a dominant feature of the modern Irish landscape. They cover about one-sixth (1.34 million ha) of the total land area (*Atlas of the Irish Rural Landscape*), which is one of the highest proportions of any European country. Perhaps because of this physical dominance, Irish boglands have been absorbed into Irish culture in a way unique in Europe, if not the world. Estyn Evans, in his classic *Irish Folk Ways*, provides detailed accounts of the tools and procedures required to extract peat and use it for fuel, but also deals with the symbolism of the peat fire in Irish society. The continuously burning fire symbolized the unbrokenness of family life and hospitality to strangers.

Figure 7.9 Bog workers, footing the turf. Robert French, Lawrence Photographic Collection. Public domain.

There is no denying the influence of bogs on the settlement patterns and transport networks of Ireland since earliest times. This influence has affected the whole rural way of life for millennia, but it is worth noting that the effect is a two-way process. The origin and development were influenced by the activities of the population. This influence is shown notably in the clearing of forests from the higher ground as agriculture developed, and by the exploitation of turf for a fuel source, particularly the industrial scale of peat cutting in the twentieth century.

For the people of Europe since the Neolithic Period, 6,000 years ago, bogs represented both a resource but also a liminal or transitional zone, a portal to the supernatural. This concept was shared by other special or sacred places such as mountain peaks, caves or springs. Around 3900 BCE, agriculture was introduced to Denmark, possibly by migrating farmers or by cultural exchange. This innovation marked the beginning of the Neolithic in the region. It was in this early phase that the practice of interring human corpses in peat bogs began. A typical example is Tollund Man, the mummified corpse of a man dated as having lived in the fourth century BCE, during the Iron Age, on the Jutland Peninsula (Figure 7.10). The cause of death was determined to be hanging, but as a result of human sacrifice rather than execution.

Currently, over 500 Iron Age bodies and skeletons, dated between 800 BCE and 200 CE, have been recovered from Danish bogs alone. Others have been unearthed in Germany, the Netherlands, Britain and Ireland. The occurrence of bog bodies in Ireland is just one aspect of the cultural significance of bogs in Ireland.

Someone who has invoked the cultural significance of bogs in the Irish landscape is the Nobel Laureate

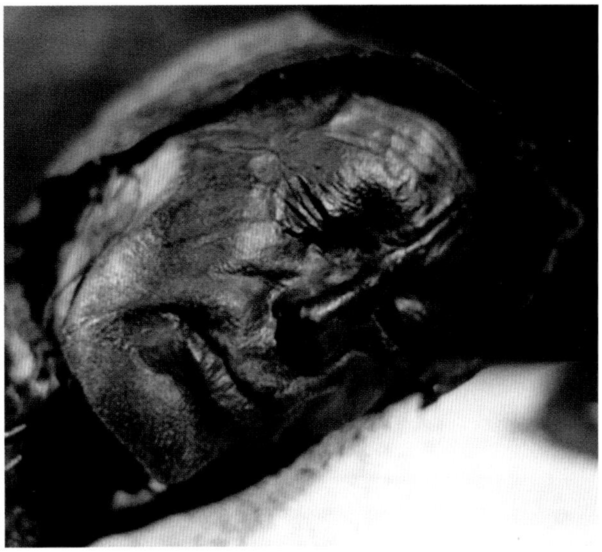

Figure 7.10 Tollund Man, bog body, Denmark, dated around 375 BCE–210 BCE. Sven Rosborn, Public domain, via Wikimedia Commons.

poet Seamus Heaney. One of his best known poems is *Digging*, in which he compares his father's ability to use a spade with his use of the pen. The poem extols the skill and stamina with which the poet's father and grandfather could cut turf (peat) from the bog.

> By God, the old man could handle a spade.
> Just like his old man.
>
> My grandfather cut more turf in a day
> Than any other man on Toner's bog.

Heaney's poem, *The Tollund Man*, reflects on human nature and the history of conflict and violence, specifically referring to the late twentieth century in Northern Ireland. The Tollund Man was ritually sacrificed and buried in the bog, supposedly for the greater benefit of his community, and this is compared to the sectarian killings in Ireland, which are also seen as sacrifices to an ideal. In both locations, these sacrifices are perceived as being ineffectual in bringing about the changes they aspire to.

In his poem, *Bogland*, Heaney describes the landscape through the metaphor of the bog – 'The wet centre is bottomless'. There are references to the Great Irish Elk, whose skeletons have been recovered from the depths of the bog, to butter buried there for preservation and found 'recovered salty and white', with the peat itself called 'kind, black butter'. The comment, 'We have no prairies/To slice a big sun at evening–', is a reference to North America, destination for so many Irish immigrants over the years, and is in contrast to the lines in the penultimate stanza – 'Our pioneers keep striking/Inwards and downwards'. The bog can be seen as a source of knowledge of the past, and this search for history can be infinite.

For many generations, the smell of peat or turf smoke was a sign of homecoming. While the decline in the use of peat as a fuel, particularly in domestic fires, means the likelihood of such an experience is increasingly rare, such is the depth of fascination of the peat bog in Irish culture that its influence on artists, poets and writers will endure. The modern recognition of the importance of bogs in the island's ecosystem means that rather than use them as a fuel source, leading to their eventual destruction, the emphasis now is on conservation and reconstruction.

8

The Landscape as a Resource

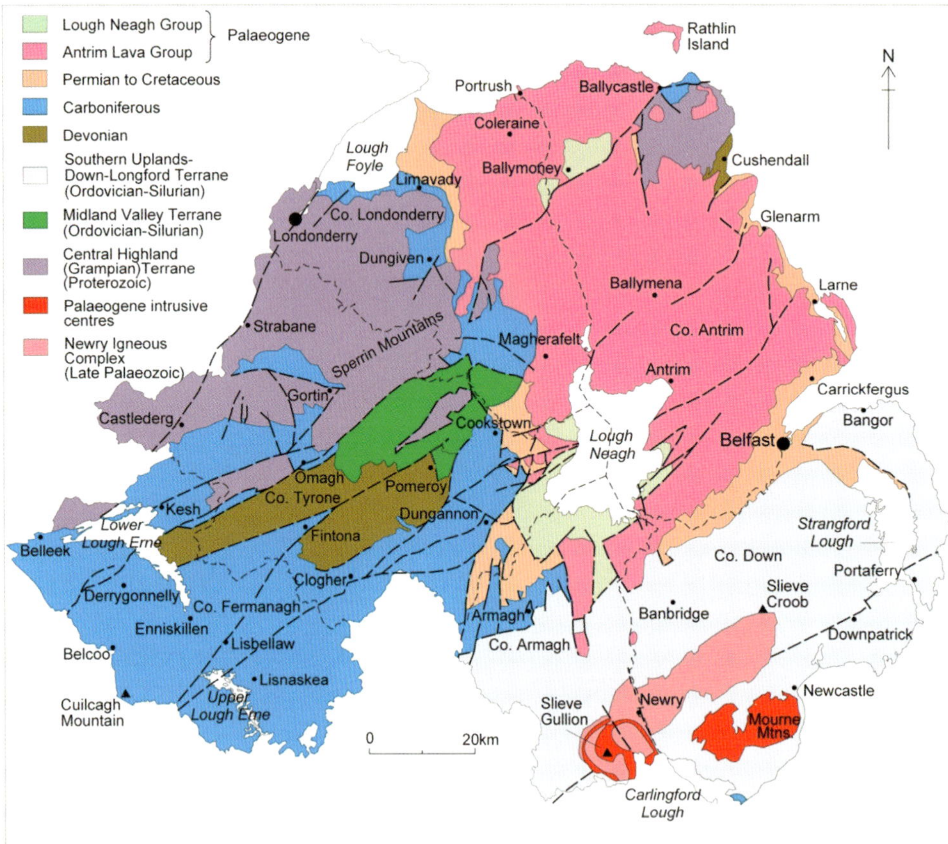

Figure 8.1A Distribution of flint outcrops in northeast Ireland. The principal occurrences of Cretaceous chalk with flint occur in a narrow strip cropping out under the lavas of the Antrim Lava Group in Counties Antrim and Londonderry. Courtesy of BGS.

Figure 8.1B Cretaceous chalk beds with flint nodules occurring in near-horizontal, darker-coloured bands, Co. Antrim.

Natural resources are taken from the landscape for a variety of reasons. These removals can be for ritualistic, artistic or aesthetic, domestic, industrial and commercial uses. In every case, however, the extraction leaves an indelible imprint on both the landscape and the branch of society that commissioned it.

Figures 1.3 and 1.4 illustrate the relationship between the natural and cultural environments, highlighting the multiple layers embedded in the landscape. These layers serve as a reminder of how the cultural environment has been shaped by the interaction between human activity and the physical landscape and influenced by evolving technologies and forms of social organization. In this context, technology refers to the extraction and use of natural resources to support the scientific and engineering advancements that accompanied increasing societal complexity – a process that ultimately culminated in the Industrial and Agricultural Revolutions of the eighteenth and nineteenth centuries.

There is a long history of exploitation of the mineral resources of the island, indeed the earliest settlers may well have been drawn to Ireland initially by the existence of that most fundamental of commodities, flint. Figure 8.1A shows the distribution of flint outcrops in northeast Ireland. Flint is a form of microcrystalline quartz occurring as small, rounded masses or *nodules* (Figure 8.1B), and its most useful properties are a hardness greater than steel and an ability to fracture in such a way as to form a razor-sharp edge. This fracturing, described as *conchoidal* because it resembles a concave shell-like shape, is exploited by flaking, or *knapping*, to produce a range of shapes such as scrapers, projectile points, knives or axes. The outcrops around the edges of the Antrim Plateau are the only *in situ* sources of flint in Ireland, although thanks to the dispersal properties of the ice sheets there is an abundance of flint *nodules* along the shore and in the near-universal boulder clay deposits. The mineral *chert* has the same chemical composition as flint, and many similar properties. In Ireland it is mainly found in the limestones of Carboniferous age, for example around Sligo and south Donegal. The early settlers used chert for the same purposes as flint.

Figure 8.2 Flint tools of Mesolithic age, Ballycastle, Co. Antrim. Notafly, CC BY-SA 3.0 https://creativecommons.org/licenses/by-sa/3.0, via Wikimedia Commons.

The short sea passage from western Scotland across the Straits of Moyle between northeast Antrim and the Mull of Kintyre is a likely route for the earliest settlers in the Mesolithic period, around 7000 years BCE. One of the earliest settlements was discovered at Mount Sandel, near Coleraine in Co. Londonderry. On arrival, these earliest colonists would have found sources of flint readily available and there is archaeological evidence for axes, hammer stones and other implements (Figure 8.2). One technique was to mount small blades or *microliths* in rows in wooden handles, which were then used for cutting or sawing.

The Manufacture of Stone Axes

During the subsequent Neolithic period (from around 4000 BCE), the advent of agriculture and the need to clear forested land for cultivation created a demand for more efficient tools than those made from simple flaked flint. This necessity drove the development of polished stone axe heads, typically fashioned from durable igneous rocks, which could maintain a sharper and longer-lasting edge. The widespread clearance of forests to enable farming marked one of the most significant transformations of the Irish landscape since the end of the Ice Age.

The study of stone axe production can be approached from a functional perspective, focusing on the materials selected and the techniques used in their manufacture. However, a deeper layer of meaning emerges when considering the significance of the quarry sites themselves. Just as the construction of megalithic monuments reflects a symbolic connection to ancestral landscapes, so too did axe quarries represent more than mere sources of raw material – they embodied a spiritual or aesthetic relationship with the Earth. In this context, the stone axe became a cultural artefact of considerable importance during the Neolithic, and its manufacture was often imbued with aesthetic and symbolic value. Numerous examples from Britain and Ireland reveal the deliberate deposition of stone axes in sacred locations, suggesting that many were produced not for practical use, but for ritual or ceremonial purposes.

A number of localities yielding suitable hard rock have been identified throughout the island, but the best-known examples were probably those exploiting a

Figure 8.3 Interbasaltic laterite horizon, Giant's Causeway, Co. Antrim.

rock called *porcellanite*, found at Tievebulliagh in north Antrim, and at Brockley on Rathlin Island, off the north Antrim coast. These locations became the sites of what have been described as axe factories, where axes were manufactured and distributed, often great distances away, in an early example of commercial activity on the island.

When basalt lava has undergone prolonged weathering it forms an iron-rich red rock called *laterite*. The laterite layer throughout Co. Antrim represents a prolonged period when the volcanic activity around Antrim was dormant and the surface lavas were weathered under a warm wet climate. Laterite is best seen in Antrim around the Giant's Causeway (Figure 8.3). Its use as a source of iron ore and bauxite (aluminium ore) will be discussed later in the chapter.

When this relatively soft material came into contact with hot magma, such as at Tievebulliagh, it was changed by the intense heat to form the very hard, dark blue rock, porcellanite. The temperature of the magma was around 1,200°C, which transformed the soft clay minerals of the laterite into hard durable minerals such as *corundum*, widely used today as an abrasive (Figures 8.4A–B). This porcellanite was capable of being highly polished and was able to hold a sharp edge, thus making it ideal for the manufacture of axes such as those found in the Malone Hoard near Belfast and now housed in the Ulster Museum (Figure 8.4C).

Axes from Tievebulliagh and Brockley on Rathlin are widely distributed across Britain and Ireland, probably an indication of their significance as ritual objects, prized for their exotic origin. The presence in

Figure 8.4A Tievebulliagh, Co. Antrim. A volcanic plug.

Figure 8.4B Porcellanite-dolerite contact at Tievebulliagh, Co. Antrim. The greenish/black rock is the hard porcellanite that was used to make axes and tools, and the brownish rock above is the dolerite. Anne Burgess, Dolerite-porcellanite Contact. https://commons.wikimedia.org/wiki/File:Dolerite-porcellanite_Contact_-_geograph.org.uk_-_472788.jpg.

Figure 8.4C Hoard of porcellanite Neolithic axes dating to 4500 BCE–2500 BCE, found at Danesfort, on the Malone Road, in Belfast. Notafly, CC BY-SA 3.0 https://creativecommons.org/licenses/by-sa/3.0, via Wikimedia Commons.

Ireland of significant numbers of axes coming from Langdale in Cumbria indicates that there was regular contact and movement of goods across the Irish Sea in Neolithic times (Cooney, 2000: 204). These items were considered exotic and of high symbolic value. There is further evidence of axe head exchanges involving sources from southwest Wales and Cornwall (see Cooney and Mandal, 1998).

In common with similar sites elsewhere in Britain, the sources of axe-making materials were considered special places, often in high or otherwise inaccessible locations, such as Pike O'Stickle in Cumbria, or on offshore islands, such as Lambay Island in Dublin Bay.

It has become increasingly evident that within Neolithic society, stone axes were far more than practical tools for woodcutting. A wide variety of rock types were employed in their manufacture, and many axes display such fine craftsmanship that they are understood to have held significant cultural or symbolic status. The sources of the raw materials used in axe production also appear to have acquired special importance within these societies.

In County Antrim, porcellanite is found at Tievebulliagh on the mainland and at Brockley on Rathlin Island, off the north coast. Recent trace element analysis of axes from both sites has revealed that the more remote and hazardous Rathlin location was used at least as frequently as the more accessible Tievebulliagh outcrop. A similar preference for challenging locations is evident at Pike O'Stickle in the English Lake District, where axes were produced from a specific volcanic rock. Durability tests have shown that equally suitable material was available at lower, safer elevations, yet was deliberately overlooked in favour of harder-to-reach outcrops. This selective sourcing suggests that the act of quarrying from 'special' places—perhaps imbued with spiritual, ancestral or aesthetic significance – was an essential component of the cultural value attached to the axes.

It is probably no coincidence that the other major Neolithic axe production site in Ireland, Lambay Island in the Irish Sea north of Dublin, was also offshore.

There, a distinctive fine-grained igneous rock known as Lambay porphyry was used to create notable stone axes.

Axes made from a variety of materials held deep cultural significance during the Neolithic, and there is strong evidence of an extensive distribution network across Ireland. Artefacts also demonstrate clear signs of two-way exchange with Britain and possibly even more distant regions. The deliberate deposition of stone axes in sacred or symbolically charged locations – commonly observed in both Ireland and Britain – further underscores their role as objects of ritual or ceremonial importance, transcending their functional utility.

Figure 8.5 Bronze Age copper mine, Ross Island, Killarney, Co. Kerry. Acknowledgement to the Geoscience Policy Division in the Department of the Environment, Climate and Communications.

The Metal Revolution

The people of the Neolithic in Ireland were the first to make significant changes to the landscape. Their predecessors in the Mesolithic period, or Middle Stone Age, around 9000 years BP, were hunter–gatherers who made limited modifications to their environment in the few thousand years during which they roamed the post-glacial landscape. In contrast, during the Neolithic, the clearance of the forest cover, the development of agriculture and the laying out of large field systems such as the Ceide Fields made profound changes in the appearance of the countryside, changes that remain with us today. The Neolithic also saw the introduction of megalithic tombs and the development of a 'ritual landscape'. These tombs often occurred in clusters, such as at Carrowmore and Carrowkeel in Co. Sligo, discussed in Chapter 5.

By the end of the Neolithic, a fairly sophisticated society must have existed, in contact with other populations outside Ireland and amenable to the introduction of new materials and technologies. The arrival of metalworking was just such an innovation and marked the beginning of the Bronze Age. See Figure 5.3 for a timeline of the early settlement of the island.

The Bronze Age (4500–2500 BP) is considered the first phase after the various subdivisions of the Stone Age and is characterized by the ability to process metallic ores. Bronze is defined as an alloy made of copper and another metal, usually tin. Ireland has a long tradition of mining for metal ores. Evidence exists for mining in Ireland dating back to around 4000 BP, when mines in southwest Ireland were producing copper at Ross Island in Co. Kerry. The mines at and around Allihies on the Beara Peninsula in Co. Cork are among the oldest workings in western Europe. The Ross Island mines are dated to around 4400–4000 BP. Geochemical analysis of the earliest surviving copper artefacts found in Ireland shows a distinctive signature of the element arsenic, and this ore can also be sourced to the Ross Island mine (see Figure 8.5).

These early Irish metallurgists demonstrated remarkable expertise in every stage of metal production – from identifying and extracting ore to refining and shaping the final product. Using simple tools such as stone hammers and shovels fashioned from animal bones, miners employed fire-setting to fracture the rock, followed by the application of lake water to rapidly cool and crack the heated stone. Once exposed, ore-bearing rock was prised from the outcrop, then crushed to prepare it for smelting.

Smelting involved heating the ore to drive off its sulphur content, leaving behind metallic copper. This molten copper was cast into bars or ingots, which were then worked by skilled metalworkers into artefacts such as axes and daggers.

At Mount Gabriel in Co. Cork, more than 30 copper mines have been located, and recent dating suggests the main period of activity was 3700–3500 years BP. Around 4000 years BP, tin was added to copper to produce bronze, a much harder alloy, and suitable for the manufacture of weapons such as axes,

Figure 8.6 The Gleninsheen gorget (gold collar), found in 1930 in the Gleninsheen region of the Burren, Co. Clare. National Museum of Ireland – Archaeology, CC BY-SA 4.0 https://creativecommons.org/licenses/by-sa/4.0, via Wikimedia Commons.

spears, halberds and domestic utensils like cauldrons. Tin ore (*cassiterite*) is not widely distributed in Ireland and there were unlikely to be sufficient reserves to satisfy the demand for bronze production in the later Bronze Age. The most likely source of tin ore was from Cornwall, possibly in exchange for copper ore or gold. The gold used to manufacture ornaments from around this period was probably sourced from alluvial deposits, that is silt, sands and gravels deposited in rivers. The Gleninsheen gorget (gold collar) (Figure 8.6) is a fine example of gold working dated to the late Bronze Age, around 2800–2700 BP, and was found in the Burren area, Co. Clare.

During the succeeding Iron Age, bog iron ore was a significant raw material. Bog iron ore, principally *limonite*, is widespread in the raised bogs of the Irish midlands and in the blanket bogs of the west. There are sparse indications of mining operations after the Bronze Age, and it was not until the late eighteenth and nineteenth centuries that the Irish mining industry revived and expanded. This was in response mainly to the demands of the Industrial Revolution in Britain. Mining for copper, lead, silver and iron was undertaken, along with extensive coal mining. Iron ore had been mined in prehistoric times and was mined during the sixteenth and seventeenth centuries, with small-scale iron workings across the eastern half of the country. The late eighteenth century also saw what could be termed a 'gold rush', with significant amounts of alluvial gold being recovered. Copper mining was important in the south and southwest of the island, with extensive

Figure 8.7 View of the 1862 engine house, Allihies, Co. Cork. The 36-foot cylinder beam engine powered the winder at Mountain Mine. Peter Bell at English Wikipedia, CC BY-SA 3.0 https://creativecommons.org/licenses/by-sa/3.0, via Wikimedia Commons.

reserves exploited at Allihies in Co. Cork. Figure 8.7 shows the remains of the nineteenth-century engine house used during the most recent phase of mining at Allihies, which began in the Bronze Age.

The Copper Coast Geopark, Tankardstown, Co. Waterford

Tankardstown, one of the principal mineralized zones along County Waterford's so-called 'Copper Coast', preserves the remains of nineteenth-century industrial heritage buildings, mineral veins and abandoned mine workings. The area's rich mineral diversity is hosted in Ordovician volcanic rocks, while the primary copper ores – probably of Devonian age – occur in veins that cross-cut the older volcanic formations.

At its height in the late nineteenth century, Tankardstown was a major centre of industrial activity. Over 1,000 men were employed in extracting ore from deep underground and in managing the constant challenge of pumping water from the mine shafts. While the majority of the workforce came from the surrounding area, some miners migrated from Cornwall and Devon, regions with long-established traditions of copper and tin mining (see Figures 8.8 A–B).

The price of copper governed the amount of copper produced by the mine. This peaked in 1865, followed by a period of fairly persistent decline, and the mine ceased production in 1879. Following the mine closure, whole families moved away, mostly to the USA, where a number of them found work at the Copper Mountain operation in Butte, Montana.

Figure 8.8A Remains of engine house and mine entrance, Tankardstown, Co. Waterford.

Figure 8.8B 'Copper Coast' cliff exposures at Tankardstown, Co. Waterford.

Mining in the North

Lead mining

At the other end of the island from the Copper Coast, from the late eighteenth century until 1900, around 13,500 tonnes of lead were produced from the mineral veins of the ten Conlig mines at Whitespots, immediately north of Newtownards in Co. Down. It was the largest mine complex in Ireland at the time and this tonnage makes it one of the most important mining districts in the United Kingdom in the nineteenth century. All that now remains of this once thriving industry is ruined buildings, overgrown spoil heaps, 'tailings' (unwanted residues of crushed ore) and capped shafts (Figure 8.9).

Figure 8.9 Lead mines, Conlig, Co. Down. Courtesy of BGS.

Figure 8.10 Adit (horizontal passage or tunnel) in the Palaeogene Interbasaltic Formation at Parkmore, Co. Antrim (Geological Survey of Northern Ireland). Courtesy of BGS.

Iron ore mining

From the mid-nineteenth century, native iron ore began to be systematically extracted around Ballycastle, Co. Antrim. By the 1860s, however, mining activity had shifted to mid-Antrim, particularly in areas such as the upper Glenravel Valley and Newtowncrommelin (Figure 8.10). Here, extraction took place in the Palaeogene Interbasaltic Formation – lateritic deposits formed by intense weathering of the Lower Basalt Formation under warm, humid conditions during a period of volcanic inactivity, prior to the eruption of the Upper Basalt Formation. The iron ore was mined using a combination of open-cast methods and underground room-and-pillar techniques.

By the early 1870s, around 700 miners were working around Parkmore, Glenravel and Newtowncrommelin, with production peaking at over 200,000 tonnes in 1880. Diminishing iron content in the ore and the high cost of transportation to Britain meant that ore from Antrim became uncompetitive. By the 1920s, ore production at Newtowncrommelin was one of the few remaining in existence.

Bauxite mining

In addition to iron ore, lateritization of the Palaeogene basalts also produces bauxite, the principal ore of aluminium. Small amounts of bauxite were recovered along with the iron ore from the Co. Antrim mines until the mid-1930s, when production ceased. However, an increased demand for aluminium for aircraft production in the Second World War led to a reopening of the mines in the 1940s, when over 200,000 tons were extracted, principally from Newtowncrommelin and Lyle's Hill, in south Antrim, near Templepatrick.

By the late nineteenth century, the depletion of known reserves and a lack of new discoveries led to a sharp decline in metal mining across Ireland. Mining activity dropped significantly, and the industry entered a prolonged slump. However, a revival began in the mid-twentieth century as the newly independent Irish state adopted policies to encourage investment in mineral exploration. This led to a resurgence of mining activity, particularly in the extraction of zinc, lead and barite during the latter half of the twentieth century.

Since the 1960s, a series of major 'base metal' discoveries have been made. Base metals – such as copper, tin and zinc – are not considered precious, but are nonetheless of significant strategic importance. The most notable of these discoveries was the large carbonate-hosted lead–zinc deposit at Navan, Co. Meath. Additional lead–zinc ores have been commercially exploited at Lisheen in Co. Tipperary and Galmoy in Co. Kilkenny. Together, these operations have established Ireland as one of the leading producers of zinc and lead in Europe.

Non-metallic resource production

In addition to metal mining, Ireland has a ready supply of industrial minerals and rocks, including coal, limestone, salt, gypsum, barite, phosphate, silica sand and dimension stone. The large demand for crushed aggregates for road and building construction has supported a thriving quarrying industry across the island. The only salt mine in Ireland occurs at Kilroot, near Carrickfergus, on the edge of Belfast Lough (Figure 8.11).

It has been in use since 1965, and the Irish Salt Mining and Exploration Company extracts as much as 30,000 tonnes of salt per week for use all over Ireland and Great Britain. The *halite* or *rock salt* occurs in beds of variable thickness within the Triassic Mercia Mudstone Group, formed by seawater evaporation in restricted basins on a shallow marine shelf. Salt extraction in the Carrickfergus area began in the mid-nineteenth century but the Kilroot mine is the only mine currently in operation. Mining at Kilroot is by the 'room-and-pillar' method, where half of the salt deposit serve as pillars to support the ceiling of the salt rooms.

Figure 8.11 Underground plant in the Kilroot salt mine, Carrickfergus, Co. Antrim. Published by kind permission of The Irish Salt Mining and Exploration Co. Ltd.

From the mid-nineteenth century until the 1950s, extraction of salt from the Carrickfergus area was by uncontrolled *brining*, where the salt is dissolved in water to produce a concentrated brine, which is then pumped to the surface. The underground caverns formed varied in size, and many have flooded since mining ceased, leading to progressive collapse of the workings and upward migration of the void. Figure 8.12 shows the surface trace caused by the underground collapse of the Tennant Salt Mine near Carrickfergus in 1990. Concentric tension cracks around the central void are clearly seen. Such collapses clearly represent a serious geohazard in an area where unregulated mining has occurred in the past, given the proximity of roads and housing to the collapse zone shown in Figure 8.12.

Figure 8.12 Surface trace of the underground collapse of the Tennant salt mine, Carrickfergus, Co. Antrim. Courtesy of BGS.

Figure 8.13A Lime kiln, Annestown Beach, Copper Coast, Co. Waterford.

Figure 8.13B Lime kiln, Ballintoy Harbour, Co. Antrim.

Limestone quarrying, lime kilns and whitewash

Lime kilns, once common throughout rural Ireland from the eighteenth century onwards, were essential structures in the production of lime. Figures 8.13 A and B show examples from opposite ends of the island. These kilns were designed to heat limestone – found in various forms across Ireland – to high temperatures in order to produce quicklime or lime dust. Quicklime had a wide range of uses in agriculture, construction and sanitation. In agriculture, lime was spread over fields to reduce soil acidity, particularly in peaty soils. To make the limestone suitable for this purpose, it needed to be reduced to small particles, a process facilitated by the lime kiln. The kiln featured a charging hole at the top for adding limestone and a draw hole at the bottom for removing the resulting powdery lime after firing. Fuelled by coal or peat, the limestone was continuously added to the kiln, and the lime produced was extracted for use in nearby fields.

Burnt lime also served as the primary ingredient in lime mortar, used for binding masonry in construction. When water is added to quicklime (calcium oxide), it transforms into slaked lime (calcium hydroxide), which, when mixed with water, forms lime wash or whitewash. This mixture was commonly used to paint walls white, contributing to the iconic image of the whitewashed Irish cottage with a thatched roof.

Coal mining in Ireland

Ireland may not have the extensive coal reserves that fuelled the Industrial Revolution in Britain, but it nevertheless has a long history of utilizing the rather more limited coal deposits it has. British coalfields had coincidental occurrence not only of coal, but also of iron ore and limestone deposits, all of these essential for steel manufacture, for example. This proximity facilitated the Industrial Revolution in Britain. While the same juxtaposition could be seen in Ireland, the coal was of an inferior quality, the seams were thinner, and the mine reserves were much smaller than in Britain.

Figure 8.14 Coal mining, Gobb Colliery, Ballycastle, Co. Antrim. Courtesy of BGS.

The first coalfield in Ireland opened in Leinster in 1638 and from the eighteenth century onwards coal was used as a domestic fuel, to smelt iron ore, to drive steam engines, and in the nineteenth century to manufacture gas for lighting and cooking. In the twentieth century, it was the dominant means of electricity generation. Coal continued to be mined in Ireland until the mid-1990s. In Northern Ireland, the Ballycastle Coalfield was the most important source before the last mine finally closed in 1967 (Figure 8.14).

The coal seams here are mainly thin seams exposed in Carboniferous sediments along the cliffs between the town of Ballycastle and the headland of Fair Head on the north Antrim coast. The method of mining was to drive *adits*, or near-horizontal shafts, southwards into the cliff face. Since the coal seams were tilted to the north, this allowed the coal to be extracted while also draining the mine. Most of the rocks in the section are sandstones and shales, with only thin coal seams of poor quality, ranging from 0.5–1.5 metres. The geological environment appears to be a subsiding coastal basin being infilled by deltaic sediments and periodically drowned by rising sea levels depositing limestones. There are three such marine beds, including the rather exotically named *McGildowney's Marine Band*.

This section of the north Antrim coast is today one of the most popular tourist destinations in Ireland, yet in the second half of the eighteenth century and early part of the nineteenth, the area was the scene of concentrated industrial activity not seen anywhere else in Ireland at the time. Much of this was driven by one man, Hugh Boyd, a local landowner.

Figure 8.15 Holy Trinity Church, Ballycastle, Co. Antrim. Andreas F. Borchert, CC BY-SA 3.0 DE https://creativecommons.org/licenses/by-sa/3.0/de/deed.en, via Wikimedia Commons.

The Boyds were a prominent family in the area, and in the 1720s, one of their members served as a colliery manager in the Ballycastle coalfield. In 1736, he acquired the rights to mine coal in the region, and over the following two decades, he expanded existing operations and opened new shafts between Ballycastle and Fair Head. To enhance the marketability of his coal, he initiated the development of several coal-dependent industries in the town.

A significant boost to this industrial expansion came in 1737, when the Irish Parliament approved the construction of a harbour. This initiative was part of a broader government strategy to facilitate the transport of coal from Irish mines – such as Ballycastle – to Dublin, thereby reducing reliance on more expensive English imports.

Mr. Boyd was, by modern standards, a visionary and entrepreneurial figure. He recognized the advantages of external market access, the availability of raw materials, a cheap and local fuel supply and a labour force eager for work. These conditions paved the way for Ballycastle's own version of the Industrial Revolution, which saw the establishment of industries such as salt production and a glassworks – both powered by coal and reliant on locally sourced sand and lime

Hugh Boyd died in 1765, and to a large extent the industries he had so assiduously built up died with him. With the exception of coal mining and quarrying, which continued on a limited scale until the late twentieth century, the other industries succumbed to a range of factors such as competition from England, a restricted local market and perhaps also a lesser entrepreneurial zeal among his descendants. Whatever the reasons, Ballycastle is now a tourist destination and the signs of Boyd's industrial revolution for the town are mostly in the realms of archaeology. Perhaps his most tangible memorial is the fine Holy Trinity Church, at the Diamond, in the centre of the town, built in 1756 of local Ballycastle sandstone (Figure 8.15).

Arigna, Co. Roscommon

Coal mining at Arigna in Co. Roscommon began in 1765 and continued until 1990 (Figure 8.16). From about 1795, the coal was used to smelt local iron ore, and the iron works continued in operation until 1838. Arigna coal fuelled Ireland's only electricity power station using indigenous coal from the late 1950s until the end of the 1980s.

The Arigna coalfield was known for its thin seams of 'culm' – a low-grade coal with a high ash content. The narrowness of the seams forced miners to work while lying on their sides, making extraction physically demanding. Initially, Arigna coal was used locally to smelt iron ore, also mined in the area. However, by the late eighteenth century, the completion of the Royal Canal enabled coal to be transported to Dublin, where it served as a cheaper alternative to imported British coal.

Figure 8.16 Coal mine at Arigna, Co. Roscommon. Gaius Cornelius, CC BY-SA 4.0 https://creativecommons.org/licenses/by-sa/4.0, via Wikimedia Commons.

The mine remained operational into the twentieth century, but when the Electricity Supply Board (ESB) power station – its primary customer – was closed in the late 1980s, the coalfield could no longer sustain operations and was shut down in 1990.

Power Generation: Water

> Night's candles are burnt out, and jocund day
> Stands tiptoe on the misty mountain tops.
>
> William Shakespeare, *Romeo and Juliet*, Act III, scene 5.

The context of Romeo's lines in *Romeo and Juliet* is his realization that the night has ended, dawn has arrived, the stars and the moon, or 'night's candles', are extinguished, and he must look to the future.

Seán Keating (1889–1977) was an Irish artist noted for his response to the Irish War of Independence and the Civil War, and the subsequent period of industrialization after the formation of the Irish Free State. In 1928–29, he painted *Night's Candles Are Burnt Out* (Figure 8.17), depicting the construction of the huge hydroelectric power station across the Shannon at Ardnacrusha in Co. Clare. The building of this major piece of infrastructure, the first project by the emergent independent state, was crucial to the idea of 'nation-building' fostered by the government and represented an acceleration of the economy and radical changes for society. The title, *Night's Candles Are Burnt Out*, can be seen as an allegory for the emergence of a new nation from its colonial history, but also has a more mundane meaning – the beginning of the electrification of the country. After this, candles are obsolete.

In the painting, Keating portrays a number of almost stereotypical players from Irish society. The drunk and the idle are confined to the sidelines under a hanging skeleton. In the centre of the picture, a man with a gun could either be conceding to the contractor or threatening him, and in the bottom right corner, a priest holding a Bible, reading by candlelight, represents the continuing presence of the church in Irish society. The dam itself is being viewed by a family – Keating himself, with his wife and two sons – as a symbol of modernity and a new era ahead.

Figure 8.17 *Night's Candles Are Burnt Out*, by Seán Keating (1928–29). Gallery Oldham, UK.

Powered by output from Ardnacrusha, the Electricity Supply Board embarked on the Rural Electrification Scheme in 1946. By the mid-1960s, this had resulted in almost every household in the Republic having access to electricity, with a parallel increase in efficiency and productivity in industry and agriculture.

The most visible impact on the landscape was the network of pylons, poles and wires seen everywhere but the most remote parts of the island. Today, the Ardnacrusha Power Station is still making an important contribution to electricity provision in Ireland (Figure 8.18).

Figure 8.18 Ardnacrusha power station, Co. Clare. The original uploader was Jcmurphy at English Wikipedia., CC BY-SA 3.0 http://creativecommons.org/licenses/by-sa/3.0/, via Wikimedia Commons.

Pump Storage, Turlough Hill, Co. Wicklow

A pumped storage scheme is a variation of the conventional hydroelectric power system, commonly used for 'load balancing' within electricity networks. It involves two reservoirs – one situated at a higher elevation, the other lower. During periods of high electricity demand, water is released from the upper reservoir to the lower one, generating power in the same way as a traditional hydroelectric system. During off-peak hours, when electricity is cheaper and demand is low, surplus energy is used to pump the water back up to the upper reservoir. This cycle allows energy to be stored and reused efficiently. In Ireland, the Electricity Supply Board established such a scheme at Turlough Hill in County Wicklow.

Here, the upper reservoir is an artificial lake constructed on the summit, the Turlough Hill Reservoir (Figure 8.19A). The workings of the power station are hidden underground in the mountain, thus reducing the visual impact of the scheme. The lower reservoir is Lough Nahanagan (Figure 8.19B), connected to the upper reservoir by an underground pipe containing the turbines for generation. The water from the upper reservoir ends up in Lough Nahanagan after moving through the station generators. That water is then pumped back up to the upper reservoir using a reversible turbine, available for generation when needed.

Figure 8.19A Upper reservoir at Turlough Hill. Albert.white at English Wikipedia. https://commons.wikimedia.org/wiki/File:Turlough_Hill.jpg.

Figure 8.19B Lough Nahanagan, the lower reservoir at Turlough Hill, Co. Wicklow. Joe King, CC BY-SA 3.0 https://creativecommons.org/licenses/by-sa/3.0, via Wikimedia Commons.

Power Generation: Wind

In the debate about the balance between traditional sources of energy, particularly *fossil fuels*, and *renewable* forms, *wind power* has been a major topic of discussion. The controversy concerning this particular mode has usually been around the siting of wind turbine generators, the noise they produce, their increasing height and the advantages and disadvantages of offshore and onshore wind farms. These recent arguments perhaps obscure the fact that leaving aside wind in sailing ships, wind power has been utilized to grind grain and pump water for 2,000 years or more. Large areas of the American mid-west, the Australian outback and the Karroo in South Africa still use wind-powered water pumps to raise water for irrigation and livestock.

Windmills first appeared in northwest Europe around the twelfth century, reaching England during that time and Ireland shortly thereafter. The earliest recorded windmill in Ireland dates to 1281 in County Wexford (*Atlas of the Irish Rural Landscape*). These early examples were post-mills – so named because the entire wooden superstructure was mounted on a central vertical post, allowing the body to rotate and face the wind. Post-mills were eventually replaced by more robust stone-built tower mills, in which the main structure remained fixed while the cap and sails rotated to catch the wind. These mills generated enough power to operate one or two millstones (see Figures 8.20A–B).

Most windmills in Ireland were located in densely populated regions along the eastern seaboard, where cereal production was most intensive (see Figure 8.21). Notable concentrations occur on coastal promontories exposed to strong winds, particularly in eastern Down and southern Wexford. Today, the stumps of these mills remain a common feature in the Irish landscape.

Mainly due to increased awareness of the long-term implications of the use of fossil fuels, there has been

Figure 8.20A Stone-built windmills from Co. Wexford. Christophe Krief, CC BY-SA 4.0 https://creativecommons.org/licenses/by-sa/4.0, via Wikimedia Commons.

Figure 8.20B Ballycopeland, Co. Down. https://www.flickr.com/photos/carolinehornet/, CC BY 2.0 https://creativecommons.org/licenses/by/2.0, via Wikimedia Commons.

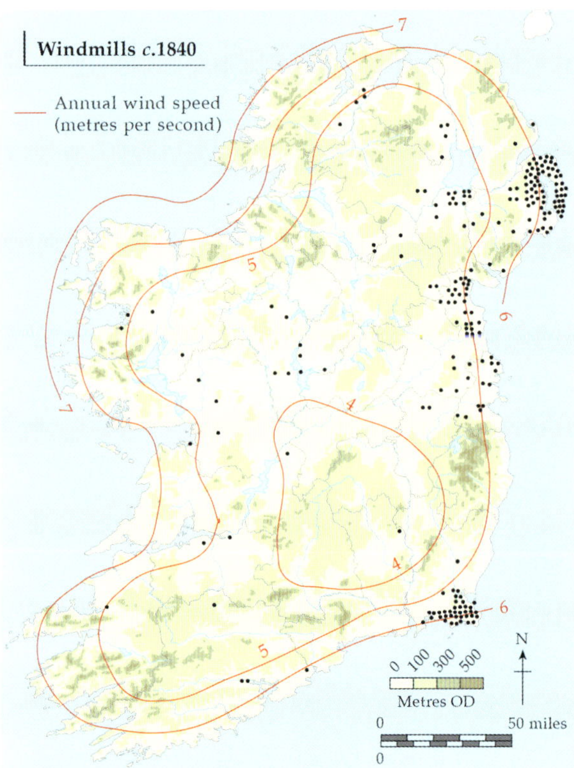

Figure 8.21 Distribution of windmills, c.1840. Courtesy of *Atlas of the Rural Irish Countryside.*

a renewed interest in the use of wind power, this time in the generation of electricity (Figure 8.22). To reduce atmospheric pollution from conventional power stations burning fossil fuels, the Alternative Energy Requirement (AER) in the Republic and the Non-Fossil Fuel Obligation (NFFO) in Northern Ireland were introduced in 1994. These require

Figure 8.22 Wind farm, Corrie Mountain, Co. Leitrim. RTG, CC BY-SA 4.0 https://creativecommons.org/licenses/by-sa/4.0, via Wikimedia Commons.

electricity distribution networks to purchase electricity from renewable sources.

In 2020, Ireland had nearly 400 wind farms and generated nearly 40 per cent of the island's demand for electricity. The environmental impact of wind turbines is a matter of continued discussion and controversy. Apart from the visual impact, which can be considerable, there are concerns over noise pollution and the impact on birdlife. Anxiety has also been expressed about the possible damage to upland areas by the provision of access roads and foundation works for turbines. These wind farms are almost exclusively onshore – to date, only the Arklow Bank offshore wind project is operational – but several projects are in the proposal stage. It is probably fair to say that offshore wind farms are as controversial as those on land.

Development of Public Water Supplies: Case Studies for Belfast and Dublin

Thick deposits of Triassic sandstone, the 'Sherwood Sandstone', underlie most of Belfast and the Lagan valley. It is porous and therefore has the ability to hold water in the spaces between the constituent sand grains. This makes it an *aquifer* and as such was an important source of drinking water in the late nineteenth century. In the days before a properly regulated public water supply, the sandstones in this aquifer were the basis of a world-famous bottled water industry that used water extracted via artesian wells. An artesian well allows water to rise to the surface without use of a pump. Companies such as W.A. Ross and Cantrell & Cochrane supplied aerated water and cordials to a worldwide market but also supplied drinking water from carts to the city residents. W.A. Ross had a deep artesian well inside their central Belfast premises and the advertising pamphlet illustrated in Figure 8.23 used a bore-hole log to show the depth of the well and thus the purity of the water. The well was around 160 metres (~500 feet) deep.

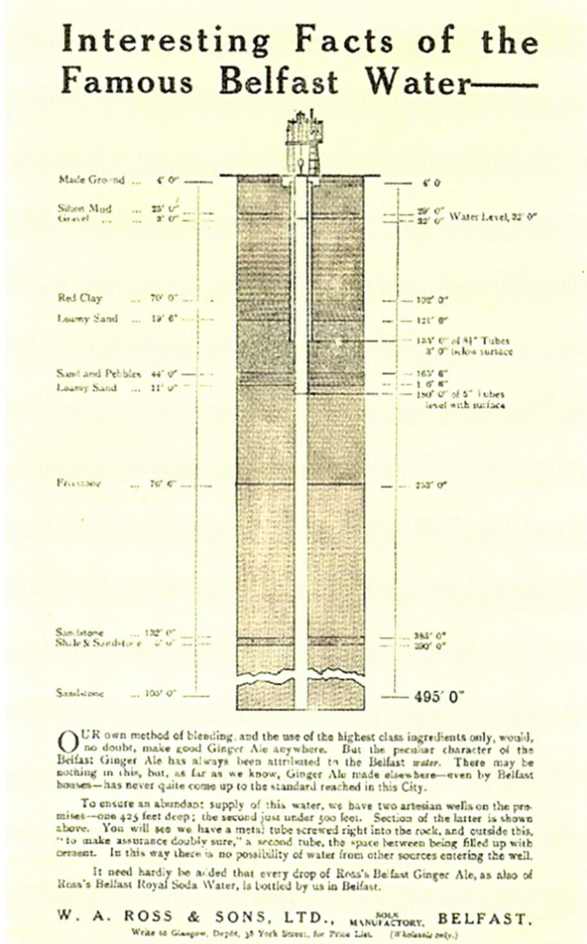

Figure 8.23 Poster advertising mineral waters supplied from the artesian borehole on the premises of W.A. Ross, in Belfast.

Building of the Silent Valley Dam

The growth of the mineral water industry in Ireland was closely linked to the rapid population expansion of Belfast during the second half of the nineteenth century and the resulting public health crises, including outbreaks of cholera, typhus, and dysentery – all associated with contaminated water supplies. As the city continued to expand, demand for clean water soon outpaced the available supply. In response, the Belfast Water Commissioners launched a major initiative in 1891 to tap the Mourne Mountains as a new source of water. A central element of this scheme was the construction of the Silent Valley Reservoir, built across the Kilkeel River valley on the southern flank of the Mourne Mountains.

The construction of the Silent Valley dam (Figure 8.24) faced significant engineering challenges. As with most earth-filled dams, a cut-off trench was required beneath the dam to prevent water from seeping underneath and compromising its stability. During excavation, engineers expected to reach solid bedrock at a depth of 15 m; however, it was not encountered until 60 metres. What had been misidentified as granite bedrock in exploratory drill cores turned out to be large glacial boulders. Due to these and other unforeseen difficulties, plans for a second reservoir in the adjacent Annalong River Valley were abandoned. Instead, between 1947 and 1951, a tunnel was constructed beneath Slieve Bignian to divert water from the Annalong River directly into the Silent Valley system. Additional small storage reservoirs, including the Ben Crom Reservoir to the north of Silent Valley, have since been built to meet growing water demand. However, increasing environmental concerns now limit the construction of new dams and reservoirs. Today, Belfast's water supply also draws from Lough Neagh – the largest freshwater lake in Britain and Ireland – which is expected to meet the city's needs for the foreseeable future.

Building of Blessington Reservoir, Dublin

Dublin in the nineteenth century shared the same problems of water supply as Belfast. The rapid enlargement of the city in the period after the Famine forced the authorities to augment supplies from the Grand and Royal Canals. Beginning in the 1860s, Dublin Corporation instigated a number of waterworks around the Wicklow Mountains to the south of the city. This included impounding the Vartry River by an earthen dam and the building of a second, higher dam, with electricity generation between the two. The Blessington Reservoir was created between 1937 and 1947 by damming the River Liffey at Poulaphouca as part of an Electricity Supply Board project to build another hydroelectric power scheme in Ireland after Ardnacrusha on the Shannon. The Blessington and Vartry reservoirs now provide the bulk of Dublin's water supply. In addition, Blessington Lake is an important area of wild bird conservation and provides important recreational facilities (Figure 8.25).

It is significant that a proposal for a further reservoir in the Mournes in the 1970s was turned down on environmental grounds. This heralded a recognition across Ireland that water is a valuable resource that must be conserved and regulated. On an island renowned for the variety of shades of green exhibited by its grass due to the abundant rainfall, there was a feeling that water was something that did not need to be paid for. At the time of writing, this notion is still very prevalent in parts, whereas in fact a modern reliable and safe water supply requires investment in infrastructure to replace pipes and conduits, which in too many places are from the Victorian era. However, the phase of dam building in upland areas in Ireland is probably at an end. Their environmental impact on the landscape is now considered unacceptable and there is a trend back to using lakes and rivers. Given the uncertainties inherent in climate change, radical fluctuations in the rainfall pattern of the island cannot be ruled out in the future and further changes in our attitudes to our water supply may be required. A factor acting against the use of loughs and rivers as a water supply is the increased incidence of algal blooms, often toxic, and caused largely by increased run-off of nitrate-rich water from farmland.

At the beginning of this book, there was a recognition that the landscape we live with was the product of the interaction between the natural environment and the

Figure 8.24 Silent Valley reservoir, Co. Down. Lofty, CC BY-SA 3.0 http://creativecommons.org/licenses/by-sa/3.0/, via Wikimedia Commons.

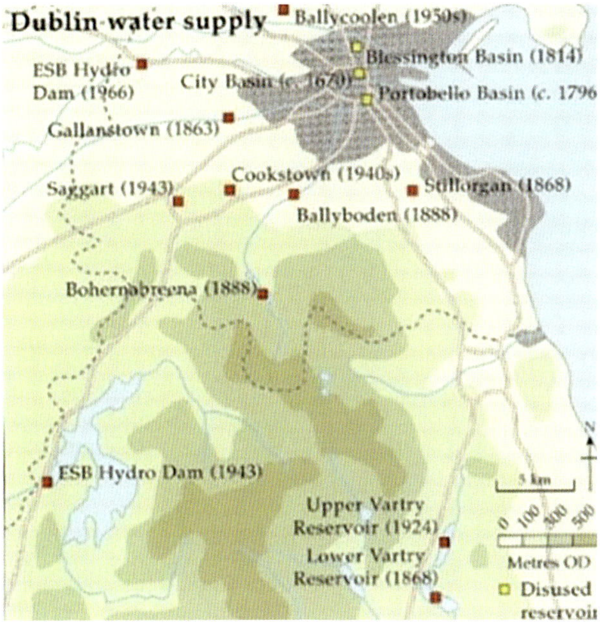

Figure 8.25 Dublin waterworks, late nineteenth century onwards. Courtesy of *Atlas of the Rural Irish Countryside*.

cultural environment (see Figure 1.4). That cultural environment included aspects of technology which were taken as the utilization of resources derived from the landscape and used to progress society. In practical terms, this ranges from the exploitation of flint nodules by the earliest settlers to the development of agriculture and the processes of industrialization we are familiar with today. Throughout this evolution, the landscape has provided the resources required to develop a rich and varied culture so well observed and documented by scholars such as Emyr Estyn Evans and Frank Mitchell (see Chapter 4).

9

The Evolution of Communications

Communications have been profoundly influenced by the Irish landscape since the earliest settlement, and in turn communications have had a profound influence on the landscape. Transport routes on land have developed from a network of paths and trackways from Neolithic times to a complex network of minor roads and laneways, which in the nineteenth century served a much greater population than currently exists on the island.

Before the Famine in the 1840s, the population of Ireland was around 8 million. In the years succeeding the Famine this number declined steeply, levelling off at around 4 million by the turn of the century before rising to its current level of around 4.8 million in the Republic of Ireland and 1.9 million in Northern Ireland. Improvements and upgrades of this pre-existing road network occurred in parallel with a system of canals, built from the late eighteenth century onwards, mainly running in an east–west direction. The Grand and Royal Canals were constructed in 1804 and 1812, respectively. These and other waterways serviced the eastern seaboard ports such as Dublin and Belfast as trade in raw materials increased with the more industrialized economy of Britain. The canals, however, did not survive the competition from the network of railways that covered the country from the 1830s onwards, but these in turn were superseded by the internal combustion engine in the first half of the twentieth century. The railway network, with its bridges, embankments and stations, had a major impact on the environment, and

Figure 9.1 Owencarrow Viaduct, abandoned railway line, Co. Donegal.

even nearly a hundred years since its peak, in the 1920s, former pieces of railway infrastructure are a common sight in both rural and urban Ireland.

The Owencarrow Viaduct near Creeslough in Co. Donegal was built in 1903 for the Londonderry and Lough Swilly Railway Company and closed in 1947. In 1925, it was the site of a derailment in which four people died. It stills forms a prominent feature in the landscape nearly eighty years after its abandonment (Figure 9.1).

The near total domination of the motor vehicle that now prevails, and the subsequent change in lifestyle to allow greater and greater commuting distances, for example, is a transport revolution that is still being played out across the country.

Paths and Trackways

The landscape that greeted the earliest settlers in Ireland around 6000 years BCE was recovering from the recent Ice Age. The climate was warming and by now the land was covered by deciduous woodlands. These earliest Mesolithic hunters and gatherers lived in forest clearings by the rivers and lakes and along the coast. Travel was along a network of paths and tracks, which would follow the line of least resistance across the landscape, avoiding poorly drained or more densely wooded areas. These routes would follow the tops of ridges and ford rivers at natural low points. By 4500 BCE, agriculture had arrived in Ireland, and Neolithic farmers were clearing the forests for grazing and cultivation, causing a profound change in the landscape over the next 2,000 years. By the end of the Bronze Age (2500 BCE), blanket bogs and raised bogs were developing, making travel more difficult. It is around this time that the earliest known trackway in Ireland was made, at Corlona Bog in Co. Leitrim. This was dated at around 1440 BCE and was found beneath 1 metre of peat, stretching for about 1 kilometre in the raised bog. It was built of planks laid longitudinally and supported by piles, and at approximately 1 metre wide it was most likely used as a footpath. Similar wooden trackways are known all over Ireland from the late Bronze Age, some wide enough to facilitate carts or wagons. A later example of such a trackway is found at Corlea, Co. Longford (Figure 9.2).

Figure 9.2 The reconstructed Corlea Trackway, Co. Longford. Kevin King. https://commons.wikimedia.org/wiki/File:Corlea_Bog_Trackway.jpg.

The Corlea Trackway is made from oak planks 3–3.5 metres long and around 15 centimetres thick, laid on rails around 1.2 metres apart. The road was at least 1 kilometre long. Dendrochronological study, whereby age can be determined from the growth ring pattern shown in the planks, suggests that the timber used in construction was felled in late 148 BCE or early in 147 BCE and the road built then, making it part of the Iron Age. It has been estimated that the sleepers alone amount to 300 large oak trees, or a thousand wagon-loads, with a similar volume of birch for the rails.

The Esker Riada

The term 'esker' is derived from the Irish *eiscir*, meaning ridge or elevation, especially one separating two plains or depressed areas. It is used to describe long, narrow, steep-sided and sinuous ridges of coarse sands and gravels deposited by meltwater streams in or under an ice sheet. In Ireland, such ridges became exposed when the ice retreated at the end of the last ice phase around 10,000 years ago and they exert an important influence on the communication pattern of Ireland, particularly in the central area, which is dominated by raised bogs.

The form of eskers is readily observed in the example from Sweden, where there is little vegetation to obscure the precise nature of the feature (Figure 9.3). The Esker Riada (Figure 9.4) is a collection of low-lying hills and ridges of sand gravel and boulders left by water flowing under a glacier during the glacial retreat around 10,000 years ago. The word 'riada' means road in Irish. The

Figure 9.3 Eskers in Sweden. Hanna Lokrantz, CC BY 2.0 https://creativecommons.org/licenses/by/2.0, via Wikimedia Commons.

Figure 9.4 Esker Riada, Co. Westmeath. Blackrock78, CC BY-SA 3.0 https://creativecommons.org/licenses/by-sa/3.0, via Wikimedia Commons.

Figure 9.5 Cross of the Scriptures, and Cathedral, Clonmacnoise, Co. Offaly. Ingo Mehling, CC BY-SA 3.0 https://creativecommons.org/licenses/by-sa/3.0, via Wikimedia Commons.

well-drained sands and gravels of the ridges contrast with the surrounding bogs and provide not only a superior transport route but also useful agricultural land. The Esker Riada still influences transport links today, as the main Dublin to Galway road, the N6, closely follows it. The Eiscir Riada is referred to in Irish as *An tSlí Mhór*, meaning the *Great Way*. The monastic settlement of Clonmacnoise was built where the River Shannon passes through it. The strategic location of the monastery made it a major centre of learning and trade by the ninth century. Largely abandoned by the end of the thirteenth century, the present-day site has a number of ruined churches, round towers, a castle and numerous carved stone crosses (Figure 9.5).

Roadways in Ancient Ireland

The *Annals of the Four Masters* (*Annála na gCeithre Máistrí*) are chronicles of medieval Irish history. They were edited and published in seven volumes between 1848 and 1851 by the scholar John O'Donovan. According to the Annals, five principal highways (*slighe*) led to Tara in early medieval Ireland:

- *Slighe Assail*: Ran westward towards Lough Owel in Co. Westmeath.
- *Slighe Midluachra:* Extended north through Slane to Emain Macha, ultimately reaching the north Antrim coast at Dunseverick.
- *Slighe Cualann*: Led south from Dublin, passing through the territory of Cualu.
- *Slighe Dála*: Took a southwesterly route through Ossory, in what is now County Kilkenny.
- *Slighe Mhór*: Known as the 'Great Highway', it joined the Esker Riada near Clonard in County Meath and followed this natural ridge westward to Galway (see Figure 9.6).

Although the Annals said the roads led to Tara, in fact they radiated out from Dublin.

It is important to recognize that this ancient road network was unlike, for example, the network of roads

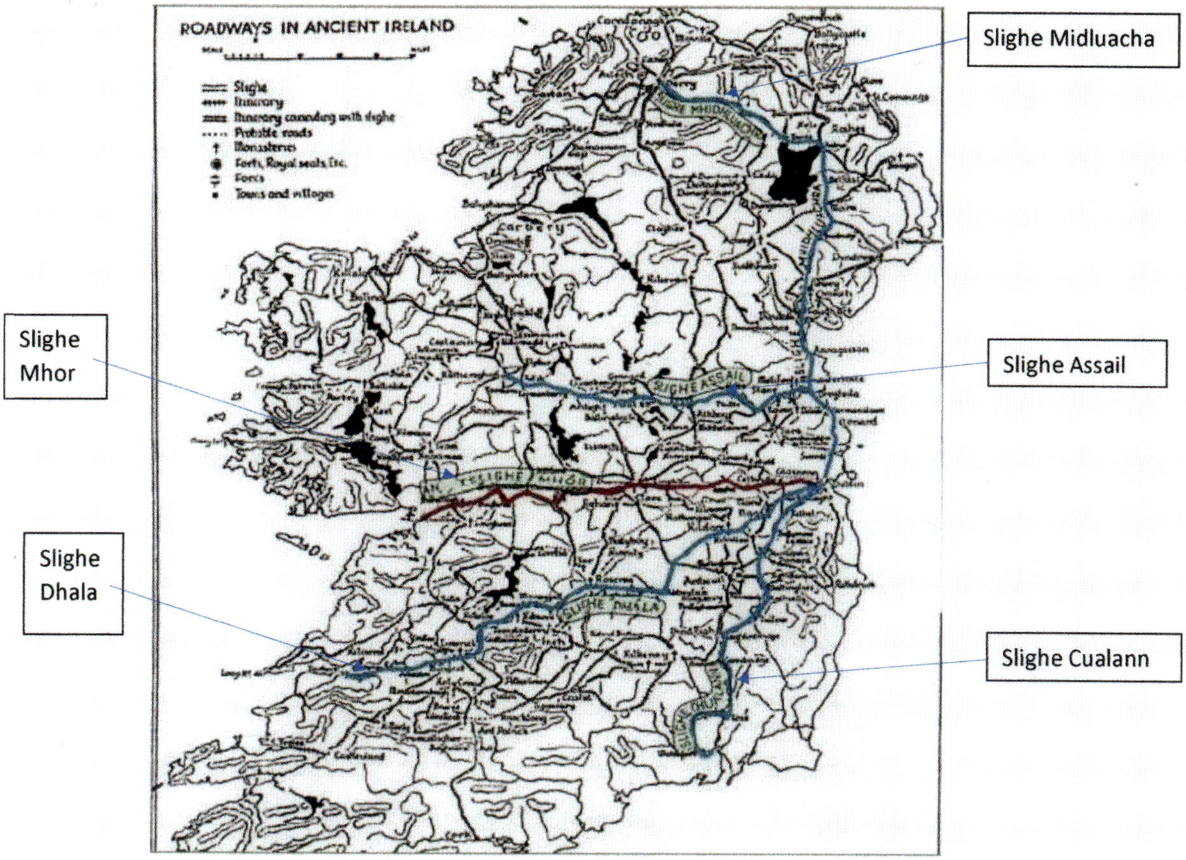

Figure 9.6 Roadways in ancient Ireland. After O'Lochlainn, Colm, Roadways in Ancient Ireland 1940. https://upload.wikimedia.org/wikipedia/commons/b/ba/.

built by the Romans in Great Britain, which were routes delineated by physical boundaries such as layers of stone pavement and parallel drainage ditches. Instead, they acted more as rights of way or traditional routes, tending to follow the line of least resistance, avoiding hazards such as bogs and poorly drained areas. As such, roads were probably not the most important transport routes up to the medieval period. Travel by sea or river was faster and easier, particularly for the movement of goods. In the sixteenth and early seventeenth centuries, efforts were made by Acts of Parliament and at parish level to improve and maintain roads within local areas, and by the mid-seventeenth century, a new Act allowed for the levying of a tax to facilitate the repair and maintenance of roads, bridges and fords. Even so, it is probably true to say that the road system as it existed was more suitable for pack animals and people on horseback or foot, rather than for any form of coach or wagon.

Towards a Modern Road Network

By the early eighteenth century, Ireland had a well-developed network of roads, the principal ones being marked on Herman Moll's *New Map of Ireland* (1714) (Figure 9.7), which showed, amongst other features, 'Passes, Bridges &c. with the Principal Roads, and the common Reputed Miles' between towns.

By the late eighteenth century, Arthur Young (see Chapter 4) was able to express his approval of the system of public roads that existed at the time. Interestingly, he contrasted the state of the 'turnpike roads' (those which had a gate or turnstile across the road, and charged a toll to pass through), which he said, 'disgraced the kingdom', with the byroads, which he described as 'admirable'. Turnpike roads were built from 1729 onwards but were never as widely used as in England, and despite their use by mail-coaches in the early nineteenth century, their use declined in the face of competition from the burgeoning railway network. They were abolished in 1858.

Figure 9.7 Herman Moll's 1732 map, showing the principal roads of Ireland. Herman Moll, CC BY 2.0 https://creativecommons.org/licenses/by/2.0, via Wikimedia Commons.

In some areas, notably Co. Wicklow, military roads were constructed in the aftermath of the 1798 rebellion. These military roads were in the same tradition as General Wade's military roads in Scotland, which were built in the Scottish Highlands in the middle part of the eighteenth century after the Jacobite rebellion of 1715. The Wicklow road was completed by 1809, by which time the insurgents had surrendered and the possibility of a French invasion, post-Waterloo had receded. The road then was incorporated into the national network and is now a popular scenic route.

By the early part of the nineteenth century, Ireland had a well-developed road network over much of the country, with the exception of a number of remote areas which would have to wait until later in the century before having roads suitable for wheeled transport. One such area was the Glens of Antrim in the northeast of the country. There were no good road links connecting the glens resulting in communities being isolated. The Board of Public Works was set up to build roads and bridges in such areas and the Antrim

Figure 9.8 Basalt over limestone cliffs, Garron Point, Antrim coast road, Co. Antrim.

Figure 9.9 Waterloo mudstone landslips at Minnis, Antrim coast road, Co. Antrim.

Coast Road was finished in 1842, providing access to settlements such as Carnlough and Cushendall. The task was accomplished by William Bald, a civil engineer who had the inspiration to build the road at sea level, along the base of the cliffs of chalk and basalt, some of which were over 100 metres high (Figure 9.8).

The specific problem with building the road at sea level stemmed from the geology of the cliffs. These consisted of basalt lava flows lying above white limestone beds, both underlain by a grey clay, the Waterloo Mudstone, which forms an unstable foundation for the cliffs. Especially during periods of heavy rainfall, the weight of the overlying basalt and limestone causes the mudstone to squeeze out, like toothpaste from a tube, resulting in major landslips that block the road (see Figure 9.9).

Bald's technique was to blast the cliff face, which then collapsed onto the foreshore and thus formed the base of the new road. The Antrim Coast Road was successfully completed in 1842, and for the next more than 100 years provided an important link between the Antrim Glens and the rest of the country. By the 1960s, however, frequent landslips meant prolonged closures of the road while repairs were made. The decision was taken to build a new road some 30 metres seaward of the old one. This required a new causeway built seawards of the original undercliff road and protected by a bank of rock armour of limestone blocks quarried locally (see Figure 9.10).

Bald created a major legacy for the communities around the Antrim glens with his inspirational vision of how a road could be made at sea level rather than inland, and in doing so overcame major difficulties that arose directly from the geological succession on the coast.

Figure 9.10 Reconstructed Antrim coast road, Co. Antrim.

Figure 9.12 Modern motorway network, showing the extent of land use involved. Seighean, Public domain, via Wikimedia Commons.

The Environmental Impact of Roads

For thousands of years, the impact of roads on the Irish environment was minimal. As noted earlier, many of the routes were suitable only for travellers on foot or with pack animals, and even when roads suitable for mail coaches were being built by the end of the eighteenth century, the designated width was 16 metres. Roadmaking was traditionally labour intensive, road building was often undertaken to provide work in distressed areas, and the materials for the road and walls were taken from the surrounding area (see Figure 9.11). This compares with the average width of a motorway in Ireland today of almost 30 metres, not including the additional land taken up by roundabouts and slip roads (see Figure 9.12).

A major environmental impact of modern-day road building is the production of aggregate required to build the road infrastructure (flyovers, bridges, etc.), as well as the various layers making up the road itself. Road making on this scale requires large quantities of crushed rock for use in bitumen and concrete mixes, which in turn requires large-scale aggregate quarries. The combined environmental impact of such quarries and the land-use demands of the motorway network is considerable. An illustration of the conflict that can exist between the needs of contemporary society and the desire to preserve the cultural heritage of the landscape was the threat to Navan Fort (*Emain Macha*), which lies to the west of Armagh City in Co. Armagh. Emain Macha is a large earthwork on top of a drumlin and was one of the earliest provincial capitals of Ulster, with archaeological evidence for complex ceremonial buildings constructed during the Bronze and Iron Ages. In the early 1980, the site was threatened by a potential expansion of a nearby limestone quarry (Figure 9.13). After a public enquiry in 1985, further quarrying was prevented and a recommendation made that the site be developed for tourism.

A similar dilemma was posed early in this century when a new motorway route was proposed that impinged on the Hill of Tara, near Skryne, Co. Meath. The Hill of Tara is an ancient ceremonial and burial site, identified traditionally as the inauguration place and seat of the High Kings of Ireland (Figure 9.14). The hill contains numerous monuments and earthworks that date from the Neolithic to the Iron Age, including a passage tomb, burial mounds, round enclosures, a standing stone and a ceremonial avenue. With such an important ritual landscape involved, there were objections that the construction of the motorway would damage or destroy archaeological evidence. Supporters of the scheme, however, argued that the motorway was an essential improvement to the road infrastructure in Co. Meath, due to its expanding population and proximity to Dublin. An additional point is that the road is a toll road that generates revenue for the state.

As with many aspects of modern life, there is a

Figure 9.11 Roadmaking in Co. Donegal, around 1900. Courtesy of the Ulster Museum.

cost–benefit analysis to be made between detrimental impacts on history and heritage and the advantage of improved transport links to the economy and the quality of life of the population.

Inland waterways and canals

Following the rapid development of a canal network in Britain around the middle of the eighteenth century, as the Industrial Revolution gained momentum, attempts were made to follow suit in Ireland. Schemes such as the Newry Ship Canal were completed by the middle of the century, along with the Coalisland Canal to service what were thought to be substantial coal reserves in Co. Tyrone.

Following on from these early schemes, other major canals were constructed during the second half of the century. In Ireland, the major trade routes were traditionally from the interior to the east and south coasts and then onwards to Britain and the rest of Europe. Ireland's longest river, the Shannon, however, flows southwest from Co. Cavan to reach the sea on the west coast at Limerick.

In the north, the River Bann flows north from Lough Neagh to the sea near Coleraine. The main ports of Belfast, Dublin and Cork are each located on shorter rivers, the Lagan, Liffey and Lee, respectively, all of which have little navigation potential. As a result of this drainage pattern, the inland waterways system that evolved was predominantly east–west. The aim was to link Belfast and Dublin with the Shannon via

Figure 9.13 Navan Fort (Emain Macha), Co. Armagh, with adjacent limestone quarry (now flooded). Copyright Crown Historic Environment Division.

Figure 9.14 Stone of Destiny at the Hill of Tara. https://commons.wikimedia.org/wiki/File:Stone_of_Destiny_2018-07-24.jpg. August Schwerdfeger, CC BY 4.0 https://creativecommons.org/licenses/by/4.0, via Wikimedia Commons.

Figure 9.15A Royal Canal, Counties Kildare, Meath and Westmeath. Silverblaster, CC BY-SA 3.0 https://creativecommons.org/licenses/by-sa/3.0, via Wikimedia Commons.

Figure 9.15B Lagan Canal, Counties Antrim and Down. This file is licensed under the Creative Commons Attribution-Share Alike 2.0 Generic license, Sue Adair.

the midlands. This was provided by the Grand Canal (completed 1805), the Royal Canal (completed 1817) and the Ballinmore and Ballyconnell Canal (completed 1859), which provided a link from Belfast to the Shannon via the Lagan and Ulster Canals (see Figures 9.15A–B and Figure 9.16).

Canals in Ireland did not share the commercial success of those in England. Part of the explanation for this was the lack of mineral resources being exploited and an over-reliance on the carriage of agricultural products. In addition, the hoped-for industrialization of the midland counties did not happen as anticipated. As a result, when the railway network began to expand

Figure 9.16 Canals and inland waterways. Courtesy of *Atlas of the Irish Rural Landscape*.

from the 1830s onwards, the canals were unable to withstand the competition.

The canal network in Ireland left significant impacts on the landscape, however. As well as the actual canal cuts, the associated development of lock-keepers' houses, warehouses, hotels and villages are still highly visible in many parts of the countryside and are an integral part of a resurgence of canal traffic for leisure purposes. A literary response to the Irish canal network was made by the poet Patrick Kavanagh. His poem, *Lines Written on a Seat on the Grand Canal, Dublin*, looks at how a memorial bench on the Grand Canal pays tribute to someone's life.

> O commemorate me where there is water,
> Canal water, preferably, so stilly
> Greeny at the heart of summer, Brother
> Commemorate me thus beautifully
> Where a lock niagarously roars
>
> Patrick Kavanagh, Copyright © Estate of Katherine Kavanagh

Railways

The age of steam in Ireland began in 1834 with the opening of the first railway line between Dublin and Kingstown (now Dún Laoghaire), designed to transport passengers from the city to the mailboats bound for Holyhead in Wales. Between 1845 and 1855, the railway network expanded rapidly, increasing from 100 to 1,000 miles of track. By 1865, that figure had doubled again, reaching 2,000 miles.

The relative efficiency of rail transport to move passengers and freight, compared with coaches or canal boats, led to a rapid expansion of the rail network during the remainder of the nineteenth century. The growth of tourism enhanced an associated boost to business as railway companies built hotels in coastal towns and encouraged the growth of day trippers to the seaside on special excursion trains. A good example of such an enterprise was the Slieve Donard Hotel in Newcastle, Co. Down, which was built in the final years of the nineteenth century by the Belfast and County Down Railway, as an 'end of line' destination, with the railway station immediately adjacent to the hotel. A spur of the railway line from the station facilitated direct deliveries of coal to the hotel.

As well as hotels and stations, railway infrastructure required substantial earthworks, tunnels, bridges and viaducts, many of which remain as prominent features of the rural and urban landscape. Figure 9.17 shows the disused railway viaduct across the Finglas River near Dingle in Co. Kerry. The viaduct carried the Tralee and Dingle Railway over the river until the closure of the line in 1953.

Figure 9.17 Finglas River and disused railway viaduct, Co. Kerry. Nigel Cox, Finglas River and disused railway viaduct. Licensed for reuse under Creative Commons Licence.

Figure 9.18 Loopline Bridge or Dublin Viaduct. YvonneM. https://commons.wikimedia.org/wiki/File:Loopline_Bridge.JPG.

The Loopline Bridge or Liffey Viaduct is a good example of railway infrastructure in an urban setting. The bridge was built around 1890 and consists of wrought iron lattice girders on a double row of piers (see Figure 9.18). Constructed as a link between north and south Dublin its claim to literary fame is its mention in James Joyce's masterpiece, *Ulysses*.

William John Leech (1881–1968) was an Irish painter whose works featured a wide range of outdoor and indoor subjects, including landscapes. He was influenced by both the impressionists like Monet and Cezanne, and by post-impressionist painters such as Matisse and Van Gogh. His painting, *Railway Embankment* (?1920 or 1938) (Figure 9.19), is an interesting study of perspective in the landscape, with the juxtaposition of the railway line, the fence and the telegraph wires, all emphasizing the landscape feature of the embankment.

The zenith of the railway network in Ireland was the 1920s, with around 3,500 miles of track across the island. It has been said that no town in the country was more than ten miles from a station. However, as with the canal network, increased competition from another transport method – this time the internal combustion

Figure 9.19 *Railway Embankment*, by William John Leech (1920). Courtesy of the Ulster Museum.

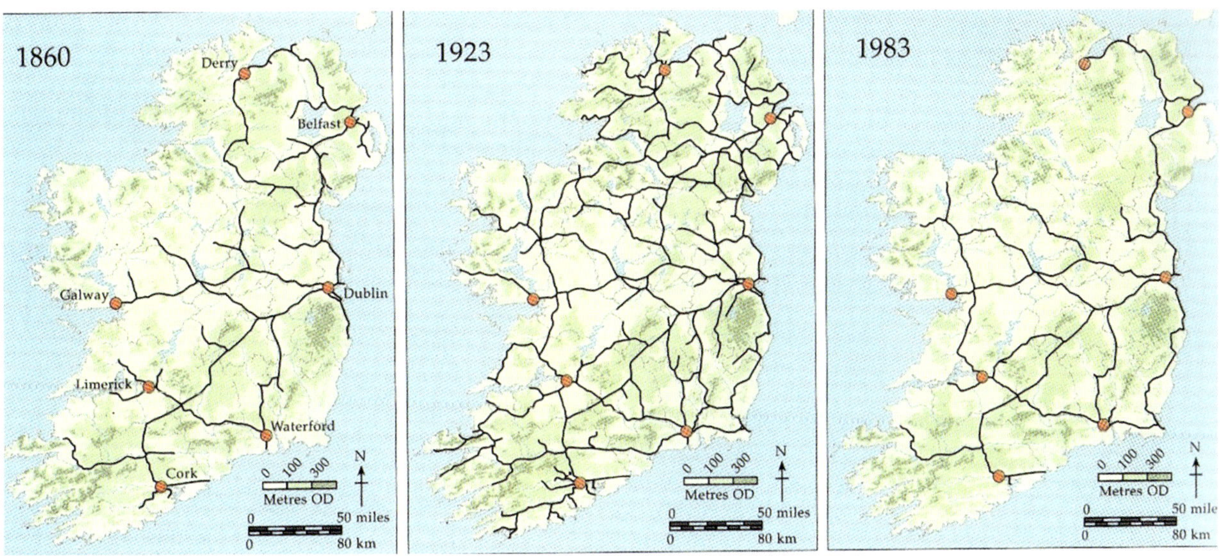

Figure 9.20 The growth and decline of the Irish railway network, 1860–1983. Courtesy of *Atlas of the Rural Irish Landscape*.

Figure 9.21A Old railway walk, Burtonport, Co. Donegal.

engine – led to a decline in railway revenues, and by the 1980s, the map of the network looked very similar to that of the 1860s, as wholesale closures took place over the entire island (Figure 9.20).

Since the large-scale decline of the railways there have been various attempts made to reuse former lines as walking or cycling routes. One such is Burtonport Old Railway Walk (Figure 9.21A), which follows part of the line that ran from Burtonport to Letterkenny in Co. Donegal. The walk has inspired a series of paintings by artist Emma Cownie (Figure 9.21B).

Figure 9.21B *The Railway Walk, Ireland.* Courtesy of Emma Cownie.

10

Landscape Evolution from the Neolithic to the Famine

The Ceide Fields (Figure 10.1) on the north coast of Co. Mayo is an archaeological site consisting of a *field system* of stone-walled enclosures. It is now known from a variety of dating methods to have been constructed in the Neolithic, around 5,500 years ago (~3500 BCE). In the 1930s, as peat was being removed from the area, it was noticed that piles of rocks uncovered suggested they had been deliberately placed and had some function. That they were below the peat layer indicated they were older than the peat and therefore of considerable age. Forty or so years later, archaeologist Seamus Caulfield, son of the original discoverer of the field system, who began to investigate the site fully, uncovered a complex of fields, houses and megalithic tombs, which had been concealed by the development of the blanket bog (see map, Figure 10.2). The subsequent excavation revealed a farming community that had cleared a wide area of pine forest and reared cattle as their main activity.

This process of clearing forest to provide arable land and building material was to continue for many centuries and was one of the profound changes in the post-glacial Irish landscape. The architecture of the field system is indicative of a social network in which the scale of the enclosures is perhaps based on household units. The scale of the building of the walls and the associated megalithic tombs is an indication of a higher level of social co-operation. How and why this particular phase of settlement ended is a matter for speculation. There may have been climatic factors coupled with changes caused by extensive deforestation, leading to social or demographic developments.

The Ceide Fields is one of the earliest recorded examples of how the landscape of Ireland has been transformed over the 10,000 years or so, since the ice retreated at the end of the last glacial period, and vegetation gradually returned. By around 10,000 BP, the entire landscape was covered by deciduous woodland, thinning out only towards the higher ground. By 6500 BP, at the start of the Neolithic and the arrival of

Figure 10.1 Ceide Fields Neolithic site, Co. Mayo. Brholden, Public domain, via Wikimedia Commons.

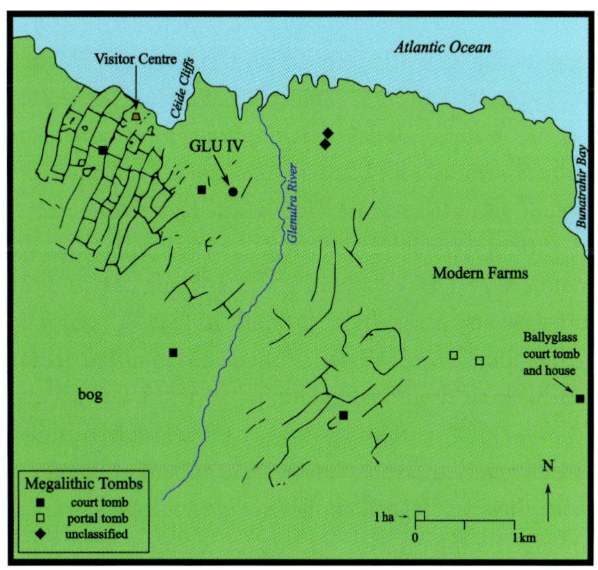

Figure 10.2 Map of the Ceide Fields archaeological site, Co. Mayo. Eroica, CC0, via Wikimedia Commons.

agriculture, farmers had begun to make clearings in the forest for grazing and cultivation. Lake basins and other hollows were progressively taken over by natural peat growth (see Chapter 7), and by 2500 BP, the end of the Bronze Age, largely owing to burning and grazing, the summit areas had been completely cleared of woodland and replaced by blanket bogs. In lowland areas, the clearance for cattle farming continued and a contemporaneous climate deterioration led to a decline in cultivation levels. By 1200 BP, population growth and an expansion in farming led to widespread environmental change, including new settlements, ringforts and associated small fields, and extensive open grazing.

From this period to the early twentieth century, the 'rundale system' was the dominant form of farming, based around nucleated groups of farmhouses known as 'clachans'. Although often referred to as villages, this term is misleading, as they lack any of the service features associated with villages, such as a church, school, shop or public house. Rundale is a form of land occupation whereby the land is divided into discontinuous plots and cultivated and lived on by a number of tenants to whom it is leased jointly.

In Scotland, a similar system existed known as 'runrig'. The 'rigs' were long strips of arable land separated by 'runs' of uncultivated land. The strips were assigned in such a way that, as in Ireland, they were scattered and would be reassigned periodically to ensure that no single tenant had perpetual use of the best land. The rundale system in Ireland had similarities to the English common field system, with the English village the nearest equivalent of the clachan. The better land was usually close to the cluster of houses and was known as the 'infield', surrounded by land which was of lower quality, often mountain land or land only useful for grazing and known as the 'outfield'. This outfield area was used for grazing in the summer months and often involved the movement of people as well as animals, a practice known as 'transhumance', or in Ireland as 'booleying'. Under this system, all the livestock grazed together, thus alleviating pressure on the crops growing in the infield area. The annual allocation of strips in the infield guaranteed an equality of distribution and was universal on the inferior lands of the West before the Famine. Seasonal migration from Ireland to England or Scotland replaced aspects of booleying in later years. In

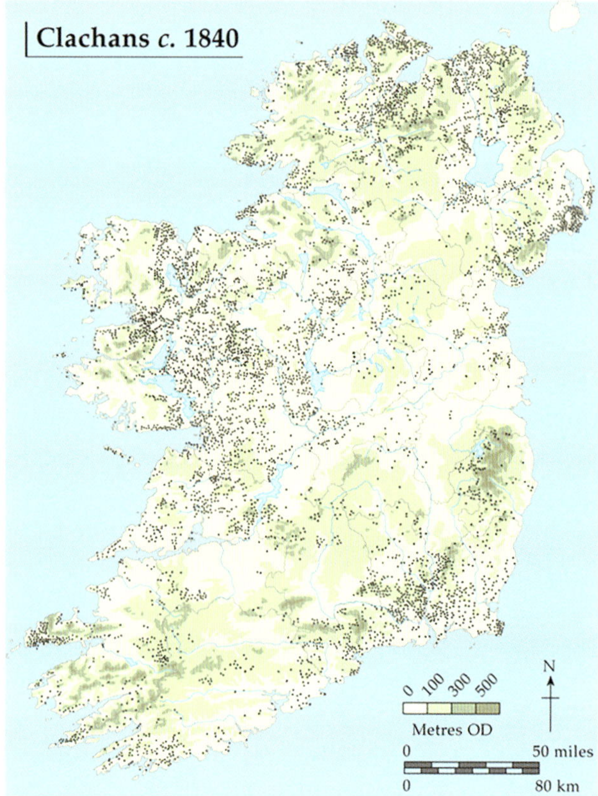

Figure 10.3 Distribution of clachans (*c.*1830) after Desmond McCourt. Courtesy of *Atlas of the Irish Rural Landscape.*

effect, the potato fields of Ayrshire subsidized the living expenses of a family for the remainder of the year. Figure 10.3 shows the distribution of clachans in Ireland by the mid-nineteenth century. Note the remarkable concentration in the west, north and southeast.

Overall, the rundale system represented a practical adaptation to specific environmental and demographic pressures. The landscape was marginal – characterized by bogs, uplands, and glacially derived soils with low fertility – while the climate was often harsh. As populations grew in these challenging areas, so did the need for effective fertilization. In the absence of artificial fertilizers, organic alternatives became essential. Seaweed emerged as a key resource, valued for its seasonal renewability and its proven benefits to potato cultivation. Rights to harvest seaweed (*cearta trá*) were highly prized, and its abundance along the Atlantic coast, combined with its agricultural utility, significantly contributed to population growth in the region.

Figure 10.4 Crocknacarra clachan, Fair Head, Co. Antrim.

Figure 10.5 Dún Mór fortified dwelling, Fair Head, Co. Antrim.

A preserved example of a clachan – a type of rural settlement associated with the rundale system – can be found at Crocknacarra, in the townland of Cross, near Fair Head in north Antrim (see Figure 10.4). This clachan consisted of three stone-built, thatched-roof houses, each typically containing two to three rooms. The residents farmed the surrounding land under leases that often spanned up to thirty years.

The immediate area around the Crocknacarra clachan shows evidence of a remarkable record of settlement over a very long period. Records exist of families living there from the early nineteenth century until the 1960s, when the last two residents were two unmarried sisters named McQuilkan (Glens of Antrim Historical Society). Currently, one building in the clachan is roofed and habitable (Figure 10.4), and the adjacent field is used as a campsite.

Cashels and Crannógs

A few hundred metres to the northwest of the Crocknacarra clachan, described above, is An Dún Mór (Great Fort) (Figure 10.5), a fortified dwelling, probably occupied from around 800–1300 CE. During the Iron Age in Ireland, hillforts or *cashels* became increasingly important. These forts served a strategic and defensive purpose and were often surrounded by earthen or stone walls.

Less than 100 kilometres to the west is the spectacular hillfort of Grianán Aileach, on top of Greenan Mountain, at Inishowen in Co. Donegal (Figure 10.6). The main structure is thought to have been built in the sixth or seventh century, although there is evidence suggesting earlier uses of the site. Dún Aengus (Dún Aonghasa), on Árainn (Figure 4.6), has some similarities with Grianán Aileach, and is also thought to have been constructed in the Iron Age.

To the northeast of the Crocknacarra clachan and Dún Mór is Lough na Cranagh, with its 'crannóg', or artificial island, constructed as a secure dwelling place for important landowners and chieftains in the area (Figure 10.7). The crannóg has remarkably well-preserved stone walls, rising in places about

Figure 10.6 Hillfort at Grianán Aileach, Co. Donegal. Silyba, CC BY-SA 4.0 https://creativecommons.org/licenses/by-sa/4.0, via Wikimedia Commons.

Figure 10.7 Crannóg, Lough na Cranagh, Fair Head, Co. Antrim. Giorgio Galeotti, CC BY 4.0 https://creativecommons.org/licenses/by/4.0, via Wikimedia Commons.

1 metre above the water level. Opinions vary as to the age of crannógs, ranging from the Iron Age to the early Christian period. This example dates back to approximately the 1600s and, according to the Ordnance Survey Memoirs of 1838, was built by one of the McDonnell chieftains who had land and a house on Fair Head (Glens of Antrim Historical Society). Crannógs occur in large numbers throughout Ireland, with more than 1,000 examples known

Field Boundaries in the Irish Landscape: Drystone Walls and Hedgerows

The field system of stone walls described earlier in the chapter, at the Ceide Fields in Co. Mayo, is merely the oldest known example of a method of delineating land boundaries by use of locally derived stones. Dry stone walls are today ubiquitous in the more rugged parts of the island, particularly in upland areas. They are seen as iconic landscape features, with a strong aesthetic appeal (Figure 10.8).

Figure 10.8 Drystone limestone wall, Inisheer, Aran Islands, Co. Galway. Carrowmarlaman: Feidin walls by louise price, CC BY-SA 2.0 https://creativecommons.org/licenses/by-sa/2.0, via Wikimedia Commons.

It has been estimated that the Irish countryside has more than 400,000 kilometres of dry stone walls (Müller, 2012). The building of dry stone walls is a cultural tradition as old as farming itself. Its origin may be explained by the need to clear the ground of large boulders in rock-strewn and glaciated environments. Rock types commonly used are sandstone or limestone, as in the example from Inisheer on the Aran Islands (Figure 10.8). The Dry Stone Wall Association of Ireland is an organization dedicated to promoting an awareness of the craft of dry stone wall building in Ireland.

In less rocky parts of the country, an alternative form of field boundary commonly used in Ireland is the hedgerow. Most Irish hedgerows were planted during the 1700s and 1800s, often in response to Acts of Parliament, which made it obligatory for landowners to erect permanent boundaries to demarcate their properties. Hedgerows planted along townland boundaries are of particular importance. Throughout the lowland farming areas, the commonest form of field boundary is the hedged ditch, faced on one side with stones and planted with a hedge on top (Figure 10.9). Commonly, enclosure was combined with drainage, with a channel at the base of the bank. The term 'ditch' is used in Ireland to include both the field bank and any associated drain or channel. It is synonymous with

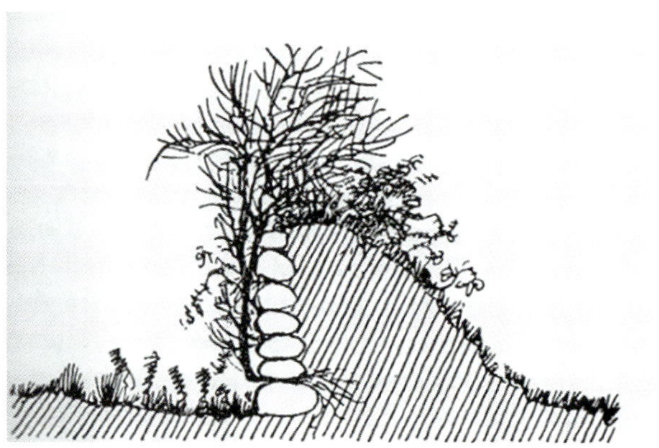

Figure 10.9. Cross-section of hedged ditch. After Estyn Evans, *Irish Folk Ways*.

the Scottish term 'dyke' but differs from the use of that word in England, where it signifies a depression. The English field hedge differs from the Irish version by being normally planted at ground level, rather than grown on a bank.

Hedgerows have important additional functions, such as creating an important wildlife habitat. They provide corridors for animal movement, nesting and roosting sites for bird species, and they provide growth sites for the native flora. In recent times, there has been a tendency to remove hedgerows as a consequence of field enlargement as farming modernized. Recognition of this as a problem has led to efforts to maintain and replace field boundaries with some limited restoration of the traditional landscape. Hedgerows are an important part of Ireland's cultural, historical and archaeological heritage. To take one example, boundary hedges for townlands are the oldest hedge types in Ireland, often dating from medieval times, and generally show a greater diversity of tree and shrub varieties. The younger hedges tend to be dominated by one or two species such as hawthorn. Field boundaries, whether as hedges or drystone walls, impart character and a sense of place to a townland or larger region, and as such are a vital component of the landscape diversity of Ireland.

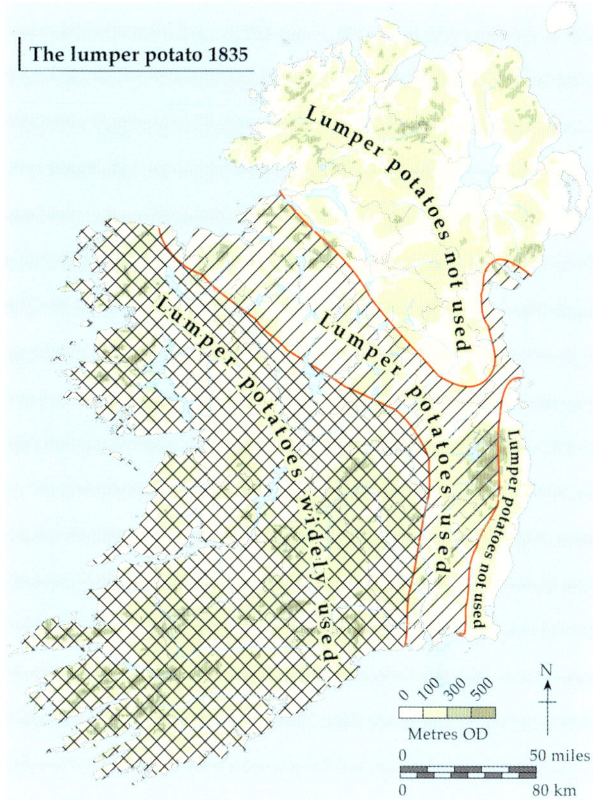

Figure 10.10 Distribution of the lumper potato. Courtesy of *Atlas of the Irish Rural Landscape*.

Potato Cultivation in Ireland and Its Social Significance

The exact history of the arrival of the potato in Ireland is obscure. There are apparently a number of contenders as to who introduced the plant, including Sir Walter Raleigh, the ill-fated Elizabethan explorer. However, there is a consensus the introduction took place in the late sixteenth century, after which time it quickly became a staple food for a substantial proportion of the Irish population. From the middle years of the eighteenth century, and for the next one hundred years or so, Ireland underwent an accelerated growth in population. Much of the dynamic of this population growth was driven by cultivation of the potato as a staple food for a high percentage of the rural population.

The cultivation of the potato in Ireland became important along the south coast in the late seventeenth century and spread rapidly through Munster as a staple part of the winter diet of the rural poor by the first half of the eighteenth century. From the beginning of the nineteenth century, there was a decline in the number of potato varieties being grown in favour of the *Lumper* variety (Figure 10.10). This variety produced high bulk yields and was tolerant of poorer soils. It quickly became widespread in Munster and Connacht and was increasingly used in Leinster by the 1830s. The main means of cultivation was by the 'lazy bed' or potato ridge method (Figure 10.11).

To construct the potato ridge, a bed around 1 metre wide and of the desired length was laid out, and the sod from alongside the bed was dug up and folded over onto the bed. This had the effect of increasing the available depth of soil and also killed any weeds growing in the sod. In *Atlas of the Irish Rural Landscape*, Aalen, Whelan and Stout (2011: 85) record the following observation: '*Ganntanas talaimh, tiubh na daoine*', which translates as 'the poorer the soil, the denser the population'. Potato cultivation saw its zenith from 1750 to 1810. It was eaten at two meals a day, while oats and milk were eliminated from the diet of the poor. Settlement had spread into

Figure 10.11 Lazy beds, Inishbofin, Co. Galway. Sheila1988,

higher and boggy areas and the combination of cheap food (the potato), cheap fuel (turf) and cheap housing (local materials), exerted a striking influence on the settlement pattern of the west of Ireland, in particular. It enabled a shift in population density from east to west, and encouraged a move into poorer land, away from the coastal areas towards bogs and mountains. Settlement moved to areas above 150–250 metres, areas of little interest to commercial farmers. This led increasingly to a potato-dependent culture, only partially ameliorated by activities such as kelp burning, cattle sales and seasonal migration. In addition, the expansion of cultivation into higher ground and wetter areas meant there was increasing susceptibility of the potato crop to disease. The overall result of these conditions was that in the west of Ireland a new cultural landscape developed allowing the population to expand enormously, from approximately 4 million in 1760 to more than 8 million by 1840 (Figure 10.12).

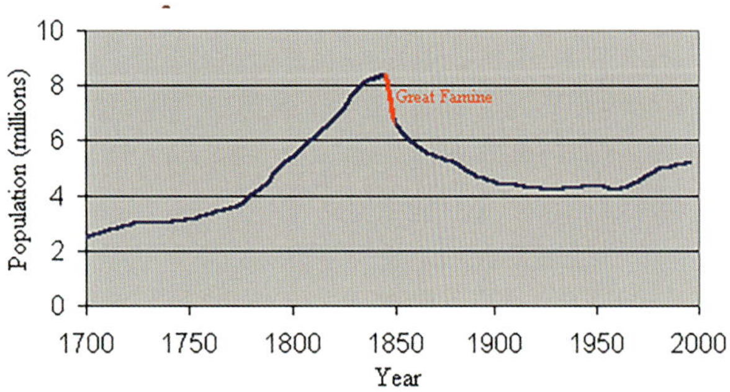

Figure 10.12 Population of Ireland, 1700–2000.

The heavy reliance on the high-yielding lumper potato reduced the genetic variety that would normally prevent the near-total destruction of an entire harvest by disease. A high proportion of the Irish population thus became vulnerable to famine. In 1845, a strain of mould, *Phytophthora infestans*, arrived in Ireland. This causes blight in potatoes, leading to wholesale rotting of the leaves and tubers of the plants. The climate that year was unusually cool and moist, conditions which allowed the blight to thrive. As a result, much of that year's potato crop was lost. This misfortune was followed in subsequent years by even more catastrophic failures, leaving millions exposed to death from malnutrition and associated diseases, such as typhus, dysentery and smallpox. As a direct consequence of the Famine, the population of Ireland fell from more than 8 million in 1844 to around 6.6 million by 1852. The fall in population numbers is accounted for by around 1 million deaths directly from the Famine, with perhaps a further 2 million people lost to the country by emigration. The impact of the Famine and its aftermath represents a defining point in Ireland's history and resulted in change in both the physical and cultural landscapes. Population numbers continued to decline for the next one hundred years, and the period resonates in the collective memory of the island and the associated diaspora to the present day.

Land Use Changes after the Famine

Significant changes in land use followed after the Famine. Notably, arable farming was eclipsed by the pastoral sector: 1.8 million hectares under arable farming in 1851 was reduced to 0.9 million hectares in 1911. There was a corresponding rise in the level of cattle exports to Britain, which was facilitated by the growth of the railway network and steamship routes. The parish of Clogheen-Burncoat is a fertile limestone corridor between sandstone outcrops of the Galtee Mountains to the north and the Knockmealdown Mountains to the south (Figure 10.13A–B).

Figure 10.13A illustrates the significant expansion of grazier farms and intensive, potato-based cultivation by 1841. This development was underpinned by a complex network of smallholdings. However, Figure 10.13B demonstrates the devastating impact of the Great

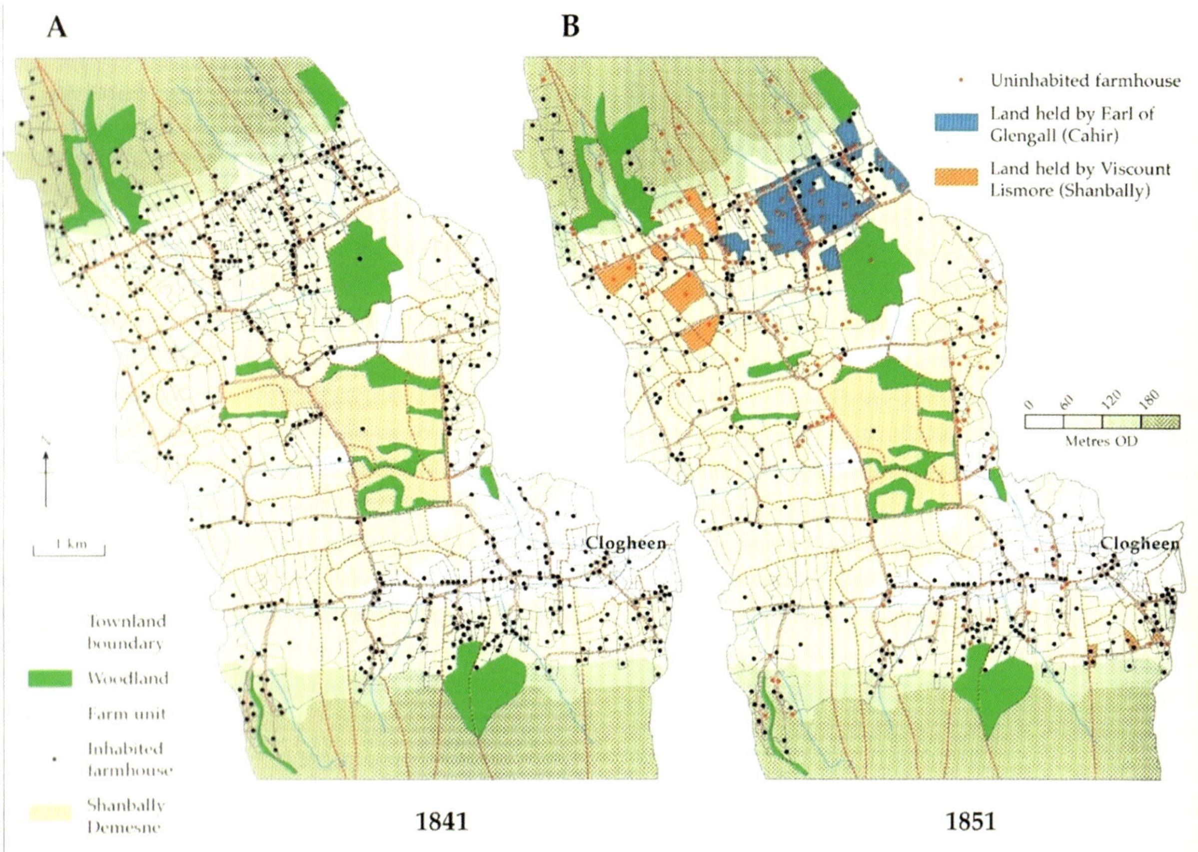

Figure 10.13A–B Pre- and post-Famine land use, parish of Clogheen-Burncoat, Co. Tipperary, 1841 and 1851. Courtesy of *Atlas of the Irish Rural Landscape*.

Famine in 1851, which revealed the vulnerability of these small farms – particularly on the northern slopes of the Galtee Mountains, where abandoned farmhouses are marked with red dots. Contributing to this collapse was the aggressive clearance of middleman tenancies by landlords, accelerating the disintegration of the existing agricultural system. Over the following fifty years, this led to a steady decline in the number of small farms and the consolidation of larger holdings. The landless population either perished during the Famine or emigrated in its aftermath.

As previously noted, the Famine had profound cultural and political consequences for the island. Most of the 2 million people who died or emigrated were Irish speakers from the most impoverished regions. As a result, the number of Irish speakers plummeted from around 4 million in 1841 to fewer than 1 million by 1891. This dramatic decline, coupled with the emergence of an English-speaking middle class, paved the way for the dominance of the English language by the end of the nineteenth century.

In addition to reshaping Ireland's cultural and political landscape, the Famine also left a lasting mark on its physical landscape. One of the most visible legacies is the network of famine roads found throughout the country. These roads were constructed through a system of forced labour, carried out under severe hardship and often inhumane conditions. According to the provisions of the Poor Law at the time, the destitute were required to perform arduous public works in exchange for minimal food relief. Among the most well-known examples is the road through the Healy Pass (R574) in Co. Kerry. Figure 10.14 shows a section of this famine road as it passes through Co. Cork.

Another enduring legacy of the Famine years is the widespread construction of workhouses, built to accommodate the vast numbers of destitute and

Figure 10.14 Healy Pass, Co. Cork. John Webster, Healy Pass on Bere Peninsula. https://commons.wikimedia.org/wiki/File:Healy_Pass_on_Bere_Peninsula_-_geograph.org.uk_-_155081.jpg, John Webster / Healy Pass on Bere Peninsula / CC BY-SA 2.0 https://creativecommons.org/licenses/by-sa/2.0, via Wikimedia Commons.

ill individuals. Following Irish independence, many of these imposing structures were repurposed as hospitals or care facilities, reflecting the evolving needs of society.

The Great Famine inflicted devastation on Ireland of an unprecedented scale, leaving profound and lasting effects on Irish society. Its consequences extended far beyond the island's borders. The mass emigration that followed gave rise to a significant Irish diaspora, particularly in the USA and Australia. Within Ireland, the sharp decline in population led to sweeping economic and demographic transformations. Over the subsequent century, the country experienced major societal shifts, including the rise – and eventual waning – of the Catholic Church's influence. The Famine's central role in Irish collective memory is evident not only within Ireland but also in diaspora communities, as reflected in the numerous memorials erected worldwide to commemorate its impact.

The Big Houses and Their Demesnes

Demesnes represent a distinctive feature of the Irish landscape, shaped over the last millennium. These estates hold considerable architectural and horticultural significance. Historically, a demesne referred to the portion of land reserved for the lord of the manor's personal use – a concept dating back to medieval times. At their height, demesnes covered over 5 per cent of Ireland's land area.

Characterized by carefully planned combinations of trees, gardens, and ornamental water features, demesnes serve as important historical records that reveal changing landscape design trends over time. By the mid-seventeenth century, estate layouts began to incorporate formal gardens adjoining the manor house. A notable feature of Irish demesnes is the presence of high enclosing walls, built to demarcate the estate and separate it from the surrounding population. These boundaries often reinforced the socio-economic divisions of the period, establishing the demesne as a distinct and privileged space (see Figure 10.15). These demesnes often included farmland, gardens, deer parks and woodlands, and at their height, they formed a prominent and carefully constructed element of the Irish landscape. Over time, they became closely – though not exclusively – associated with the Protestant Ascendancy in Ireland. The owners of these estates were typically part of the landed gentry, a social class that was later expanded to include individuals from professional, business and merchant backgrounds who aspired to the same status.

From the eighteenth century onward, a flourishing of landscape design ideas transformed these estates. Influenced by continental European trends, estate owners introduced features such as sweeping, tree-lined avenues, ornamental lakes, exotic or walled gardens, boathouses and architectural follies. These follies were decorative structures, often inspired by classical motifs, built purely

Figure 10.15 Boundary wall, Birr Castle Demesne, Co. Tipperary. Boundary Wall, Birr Demesne by Jonathan Thacker, CC BY-SA 2.0 https://creativecommons.org/licenses/by-sa/2.0, via Wikimedia Commons.

Figure 10.16 Mussenden Temple, Downhill, Castlerock, Co. Londonderry. D LN, CC BY-SA 4.0 https://creativecommons.org/licenses/by-sa/4.0, via Wikimedia Commons.

for aesthetic purposes and to enhance the visual drama of the estate. The construction of such features was sometimes the result of eccentric personal tastes. One notable example is Frederic Hervey, Bishop of Derry and Earl of Bristol, whose distinctive folly stands atop the cliff overlooking Downhill Strand (Figure 10.16).

Constructed in the late eighteenth century and inspired by the Temple of Vesta in Rome, the structure was originally surrounded by sufficient land to allow a carriage to circle the building. However, due to ongoing coastal erosion, much of the surrounding ground has since disappeared, leaving the temple perilously close to the cliff edge.

Figure 10.17 depicts Emo Court, once the largest country estate in Ireland. Designed by James Gandon – the renowned architect behind Dublin's Custom House and Four Courts – Emo Court was commissioned in the final decade of the eighteenth century for the Earl of Portarlington. It exemplifies the grandeur of a demesne estate, featuring formal gardens, woodland walks, a 20-acre lake, and an array of classical statues that reflect the aesthetic sensibilities of the period.

The young Earl of Mornington—later known as the Duke of Wellington, famed for his victory at the Battle of Waterloo—received part of his education in Portarlington. The demesne's main approach was once marked by two striking parallel rows of redwood trees (*Sequoia*, genus *Wellingtonia*), a botanical tribute to his legacy. In 1994, the then-owner, Major Cholmeley-Harrison, donated Emo Court and its surrounding estate to the nation, placing it under the care of the Office of Public Works.

The romanticized vision of the Irish landscape, exemplified by the development of demesnes and grand houses, was a theme embraced by many artists of the era. One such example is William Ashford's painting of Mount Kennedy in County Wicklow – a two-storey Georgian house completed in the 1780s – shown in Figure 10.18. Ashford, one of Ireland's most esteemed landscape painters, was known for his topographical views of country estates, often commissioned by members of the landed gentry.

Figure 10.17 Emo Court, Co. Laois. AnTuisealGinideach, CC BY-SA 4.0 https://creativecommons.org/licenses/by-sa/4.0, via Wikimedia Commons.

Figure 10.18 *Mount Kennedy, Co. Wicklow*, by William Ashford (1746–1824). Public domain, via Wikimedia Commons.

Figure 10.19 *An Allegory*, by Sean Keating (1924). Courtesy of National Gallery of Ireland.

Yet the landscape is also marked by the ruins of numerous derelict Big Houses, scattered throughout the countryside. Their decline can be traced in part to the sweeping land reforms introduced by the Irish Free State after independence in 1922. These Land Acts aimed to dismantle the estate system and redistribute land ownership. Another major factor was the burning of many of these estates by the IRA during the War of Independence (1919–21), leading to the destruction or abandonment of a once-dominant class of residences.

The work of artist Sean Keating was examined in an earlier chapter when discussing power generation in Ireland. His evocative picture, *An Allegory* (Figure 10.19), is taken as comment on the aftermath of the Irish War of Independence and the subsequent Civil War.

The painting, which was first exhibited in 1924, shows the two opposing sides in the Irish Civil War, one wearing the uniform of the newly formed *Free State National Army*, the other more casually dressed as an *Irregular*, burying a coffin draped in the Irish tricolour, at the base of an ancient tree. The well-dressed businessman and the priest talking conspiratorially represent establishment interests, and the mother and baby are Mother Ireland and future generations. In the background can be seen the ruins of an eighteenth-century Big House in spacious grounds, a standard signal for the damage inflicted by both sides in warfare. It also serves as a reminder of Ireland's Ascendancy past and the change in the old order of landed gentry and tenant farmers. The Land Acts introduced by the Irish Free State brought about a significant redistribution of land from large, untenanted estates to individual farmers. As a result, land ownership among Irish farmers rose dramatically – from just 3 per cent in 1870 to 97 per cent by 1929. *An Allegory* can be interpreted as a powerful critique of war in general, and of the Irish Civil War in particular. It suggests that Irish society is engaged in the destruction of both its historical legacy and its nationalist ideals. At the centre of the composition, the slumped figure serves as a self-portrait of the artist, reflecting his sense that the idealism once driving the struggle for independence had been compromised by disillusionment and the pursuit of self-interest

The Demesne in Twentieth-Century Ireland

The cumulative impact of the various Land Acts enacted in the late nineteenth and early twentieth centuries was profound. The compulsory purchase and redistribution of parkland led to the subdivision of over 100,000 hectares into farmland or afforestation. While some of the Big Houses found new life as schools, hospitals, convents or hotels, many more were either demolished or left to decay – adding to the ruins already left in the wake of the Civil War.

By the latter half of the twentieth century, however, public attitudes began to shift significantly. There was growing recognition of the cultural, architectural and ecological value of demesnes within the Irish landscape. Today, many of these houses and their surrounding estates – on both sides of the Border – are under public ownership and stewardship. In the Republic of Ireland, they are managed by state agencies such as the Office of Public Works, while in Northern Ireland, organisations like the National Trust have taken on a key role in their preservation.

Nonetheless, the future of demesnes remains complex. As the landscape continues to evolve alongside societal needs, pressures from modern developments – such as the construction of golf resorts to support the tourism sector – pose ongoing challenges to their conservation and integrity.

11

Castles and Cloisters

Castles

The Norman invasion of Ireland began in 1169 at the invitation of Dermot MacMurrough, the deposed King of Leinster. Seeking to reclaim his throne, MacMurrough enlisted the help of Norman forces, offering loyalty to King Henry II of England and promising land in return. This intervention marked a turning point in Irish history, setting in motion significant cultural and political changes. One of the most visible and enduring legacies of the Normans was their transformation of the Irish landscape through castle building (see Figure 11.1).

To secure their foothold in Ireland, the Normans recognized the strategic necessity of constructing fortified structures. As in England and Wales, their earliest castles were mottes – artificial, flat-topped mounds on which wooden or stone keeps were built. These were typically accompanied by a bailey, a fortified courtyard adjacent to the motte. The distribution of motte castles across Ireland closely mirrors the Anglo-Norman pattern of conquest, with a marked concentration in the eastern regions. Of the approximately 350 mottes across the island, about 180 are

Figure 11.1 Dunlough Castle Co. Cork. Wikipedia credit. Dunlough Castle Co. Cork is one of the oldest in Ireland and characterized by dry stone masonry building without the use of mortar between the stone blocks. https://commons.wikimedia.org/wiki/File:2019-07-30-Dunlough_Castle-0789.jpg, © Superbass / CC BY-SA 4.0 (via Wikimedia Commons), CC BY-SA 4.0 https://creativecommons.org/licenses/by-sa/4.0, via Wikimedia Commons.

Figure 11.2A Clough Castle motte and bailey, Co. Down. Ardfern, CC BY-SA 3.0 https://creativecommons.org/licenses/by-sa/3.0, via Wikimedia Commons.

Figure 11.2B Knockgraffon Motte, New Inn, Co. Tipperary. Sarah777, Public domain, via Wikimedia Commons.

found in Leinster and around 120 in Ulster (Figures 11.2A–B).

Probably the most imposing example of Norman castle building is Trim Castle in Co. Meath, constructed by Hugh de Lacy. This imposing fortress featured a curtain wall, moat and keep, and its current form was completed after de Lacy's death in 1186 (Figure 11.3). Soon after, other significant Norman castles followed: John de Courcy built Carrickfergus Castle in Co. Antrim, and Theobald Walter, First Baron Butler, established Nenagh Castle, distinguished by its iconic round tower. Castle construction by the Normans continued into the thirteenth century, with prominent examples including Dundrum Castle in Co. Down and both Carlingford and Greencastle Castles, strategically positioned on opposite shores of Carlingford Lough on Ireland's east coast.

During the fourteenth century, Norman influence in Ireland began to wane, undermined by a series of external and internal pressures. The Great European Famine of 1315–17, followed by the devastating Black Death in the late 1340s, severely weakened their settlements. At the same time, resurgent Irish kings and chieftains had adapted their military tactics, becoming more effective in resisting Norman dominance. The overall effect was a period of economic stagnation and political decline. As a result, the large-scale construction of castles largely ceased during this era, giving way to a new architectural form: the tower house.

Tower houses began to emerge in the early fifteenth century and continued to be built into the seventeenth century. Over 200 examples still survive today,

Figure 11.3 The keep and curtain walls, Trim Castle, Co. Meath. Andrew Parnell, CC BY 2.0 https://creativecommons.org/licenses/by/2.0, via Wikimedia Commons.

Figure 11.4 Tower house with bawn, Aughnanure Castle, Oughterard, Co. Galway. Christophe Meneboeuf, CC BY-SA 3.0 https://creativecommons.org/licenses/by-sa/3.0, via Wikimedia Commons.

reflecting their widespread adoption across the island. These fortified residences were constructed by both the Anglo-Irish and Gaelic Irish, as well as by English and Scottish settlers during the plantation periods of the late sixteenth and seventeenth centuries. Designed for both residential and defensive purposes, many tower houses were enclosed by a defensive wall or *bawn*. A notable example is Aughnanure Castle in Co. Galway, built in the sixteenth century by the O'Flaherty clan (Figure 11.4).

Castles, including their later adaptations such as tower houses and fortified manor houses, have left a lasting imprint on the Irish landscape. Strategically located in both rural and coastal settings, they were constructed to assert authority and provide protection during times of frequent conflict and invasion. While the most iconic examples stem from the initial wave of Norman stone castle building in the twelfth century, the decline of Norman control saw a shift toward native Irish fortification, culminating in the tower house as a dominant architectural response to the turbulent conditions of the late medieval and early modern periods.

Cloisters

In the Introduction, I discussed land use around Strangford Lough in Co. Down including a description of the monastic settlements at Greyabbey and Nendrum (see Figures 1.8 and 1.9). Greyabbey, founded in 1193

Figure 11.5 Round tower, Glendalough, Co. Wicklow. Superbass, CC BY-SA 3.0 https://creativecommons.org/licenses/by-sa/3.0, via Wikimedia Commons.

by the wife of John de Courcy – an Anglo-Norman invader of Ireland and the builder of Carrickfergus Castle – reflects the Norman tradition of establishing religious sites. In contrast, Nendrum was an early Christian monastery, dating back to around the seventh century, and forms part of Ulster's pre-Norman ecclesiastical heritage.

The Norman foundation of monasteries continued a long-standing tradition of monastic construction in Ireland, which began with the arrival of Christianity in the fifth century. These monastic centres served not only as places of worship but also as hubs of education and learning. Christianity brought about significant societal changes, including the advent of literacy, manuscript production and the emergence of a distinctive ecclesiastical architecture.

The early Christian period left a lasting imprint on the Irish landscape, evident in the remains of monasteries, abbeys, carved stone crosses, shrines and round towers (see Figure 11.5). The round tower, or *cloigtheach* (meaning 'bell house'), is a striking feature of many early monastic sites. Most were built between the seventh and tenth centuries, rising up to 34 metres in height, with entrances typically positioned 1.5 to 4.5 metres above ground level. While these elevated doorways are often interpreted as a defensive feature – allowing monks to seek refuge during Viking raids – they may also have contributed to structural stability by reducing pressure on the tower's base.

Construction typically involved two closely spaced concentric walls filled with rubble – a technique inherited from Roman engineering. The presence of *entasis* – a slight convex curvature in the tower's shaft designed to correct visual distortion, a hallmark of classical architecture – suggests that Irish monastic builders may have adopted building techniques introduced by missionaries from England or continental Europe.

The Rock of Cashel

The Rock of Cashel serves as an exemplar of the legacy of monastic building in medieval times on the landscape of modern Ireland. With a spectacular setting on an outcrop of Carboniferous limestone in the Golden Vale of Tipperary, the Rock has the most impressive grouping of medieval buildings to be seen anywhere

Figure 11.6 Rock of Cashel, Co. Tipperary. Irish Typepad, CC BY-SA 2.0 https://creativecommons.org/licenses/by-sa/2.0, via Wikimedia Commons.

in Ireland. On the site, among other features, there is a round tower, a high cross, a Romanesque chapel, a Gothic Cathedral, an abbey, the Hall of the Vicars Choral and a fifteenth-century Tower House. Little wonder it is one of the most popular tourist attractions in Ireland (Figure 11.6).

In a similar way that castles were used to dominate the landscape and impose control and authority over an area, so churches and other ecclesiastical buildings have been used as a means of asserting ownership of the landscape and to influence the interpretation of space.

12

Epilogue: Whither the Landscape?

The landscape, like language, is constantly evolving. 'Whither', a literary or old-fashioned word meaning 'to where' or 'to what place or state', seems to me to be an appropriate word to use about the future condition of the Irish landscape.

The scenery we are familiar with today on the island of Ireland is the result of the amalgamation of a range of factors that include the composition and distribution of the bedrock and the workings of weathering and erosion since the retreat of the ice some 13,000 years ago. Superimposed on these natural processes are the cultural changes brought about by the inhabitants of the island since their arrival on these shores about 8,000 years ago. These changes reflect the advances in technology and the changes in society as agriculture developed and the Industrial Revolution took hold, leading to the highly developed society most of us live in today.

What is immediately evident is that the rate of environmental change has accelerated dramatically over the past few centuries. The earliest settlers in Ireland, probably arriving during the Mesolithic period around 8000 years BP, lived as hunter–gatherers for several millennia and had minimal impact on their surroundings. The advent of agriculture marked a turning point, initiating the clearance of woodland for cultivation and grazing, and contributing to the formation of peat bogs – changes that profoundly reshaped the landscape and set the stage for the environment we see today.

Population growth was gradual but steady, eventually reaching over 8 million by the early nineteenth century. This population was heavily dependent on a single staple – the potato. When successive harvests failed in the 1840s due to potato blight, the result was catastrophic: widespread famine, significant loss of life and mass emigration. The Great Famine left a deep imprint on both the physical and cultural landscape. The transformations that followed in its aftermath mark the beginning of modern Ireland, laying the foundations of the society in which we now live.

A useful indicator of the sort of changes taking place, and their rate of change, is the level of the atmospheric gas carbon dioxide (CO_2). Carbon dioxide is an important heat-trapping (greenhouse) gas, which is released through human activities such as deforestation and burning fossil fuels, as well as natural processes such as respiration and volcanic eruptions.

Analysis of ancient air bubbles trapped in ice enables climate scientists to see what the Earth's climate and atmosphere were like in the past. This shows that levels of CO_2 in the atmosphere are greater than they have been at any time in the last 400,000 years. During glacial periods, atmospheric CO_2 levels were approximately 200 parts per million (ppm), rising to around 280 ppm during warmer interglacial periods. By 2013, CO_2 concentrations surpassed 400 ppm for the first time in recorded history – a significant milestone. This sharp increase is closely linked to the widespread burning of fossil fuels that began with the Industrial Revolution around the mid-nineteenth century, when CO_2 levels were approximately 280 ppm. Today, atmospheric CO_2 has reached concentrations roughly 50 per cent higher than pre-industrial levels, reflecting the profound impact of human activity on the global climate system.

While geologists have declined to formally designate a new epoch called the Anthropocene, there is little doubt that human activity is the main factor now in environmental change. Over the past 170 years or so, human activities have raised atmospheric concentrations of CO_2 by 48 per cent above pre-industrial

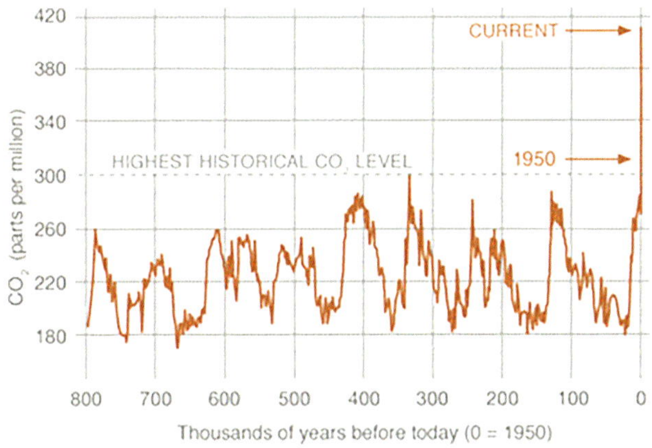

Figure 12.1 Historical levels of atmospheric CO_2 levels, 800 kya to present: reconstruction from ice cores. NASA.gov.

levels, starting in 1850 (Figure 12.1). To put this in the context of a longer timescale, it is more than what had happened naturally over the 20,000-year period from the Last Glacial Maximum (the most recent time that ice sheets were at their greatest extent) to 1850, a rise from 185 ppm to 280 ppm). These changes, which are global in extent, have a number of impacts that are specific to Ireland and its landscape. Changes expected include temperature increases of around 2° C by the end of the century, changes in the rainfall pattern, with possible increases in winter and decreased values in summer. The extent of climate change is open to speculation, Ireland has little influence over other larger nations and their energy policies and there is a great deal of uncertainty over the political will required to make the changes required to reduce energy consumption and the attendant greenhouse gas emissions.

What is certain, however, is that what we recognize as the culture of Ireland in the broadest sense is closely bound up with the landscape in its entirety, and that literature, poetry, the visual arts, music and aesthetic appreciation of the countryside in all their manifestations are defined by that connection. Irrespective of what transpires in the next few decades elsewhere among the nations of the world, it is already possible to recognize a number of major threats to the Irish landscape and therefore by projection to the Irish people.

Threats directly to the landscape include the increased frequency of structures appearing in the landscape such as wind turbines, electricity pylons and telecommunication masts. Large numbers of wind turbines or solar panels such as are advocated as renewable energy sources, whether sited onshore or offshore, if unsympathetically located, can constitute a major blight. In addition, wind turbines located in upland areas have already caused serious problems when bog drainage has been altered during foundation works for the towers, which can be more than 100 metres high.

Further potential threats relate to factors such as:

- increased urbanization/suburbanization;
- agricultural activities, including aquaculture, such as fish and shellfish farming;
- gas and oil pipelines;
- waste disposal, particularly landfill operations;
- water-quality issues – pollution of groundwater, including fracking for gas retrieval, surface water contamination, marine environment pollution;
- mining activities, including tailings ponds, spoil heaps, remediation issues;
- quarrying activities;
- electricity generation including hydroelectric power projects, pump-storage schemes, tidal energy schemes;
- Landscape changes due to large-scale transport infra-structure projects.

The change of policy by Bord na Móna from peat bog exploitation to conservation represents a fundamental reversal from the industrial removal of peat to feed power stations to their preservation in the landscape. Since this is a relatively recent development, the full implications have still to be understood, but clearly the abolition of peat removal and the potential for re-wetting drained bog areas means that wide areas of the midlands need remediation. Bord na Móna has announced the biggest change of use involving Irish land in modern history. Some 125,000 acres (~51,000 hectares) of bogland that are now being used to provide energy peat to three power stations will transition to new uses by 2030. After 2030, the company will no longer harvest energy peat and will have completed its move to new sustainable businesses, located across its bogs and landholding. The move will involve the rehabilitation of tens of thousands of acres of Irish bogs, providing new biodiverse habitats that can also support eco-tourism and community amenity resources. Other developments will include renewable energy generation, resource recovery and horticulture.

In recent years, there has been an increase in awareness among the general public of the importance of the Irish landscape as a resource to be protected and nurtured. This is to the benefit not only of the indigenous population, but also to the large numbers of visitors to the island who come in search of a unique cultural experience. There is now a greater recognition of the social and cultural value of the landscape, but also of the economic benefit to be derived from it. This benefit derives not just from visitors coming into the country, but also from residents recognizing, perhaps for the first time, the quality of the asset on their doorstep. At the time of writing, one of the effects of a worldwide pandemic has been a restriction in foreign travel that has engendered the concept of a 'staycation' – a holiday in one's own country. It may be that in the aftermath of the pandemic such an idea takes hold permanently.

Select Bibliography and Further Reading

Aalen, F. H. A., Whelan, K. and Stout, M. (eds) (2011) *Atlas of the Irish Rural Landscape*, 2nd ed. Cork: Cork University Press.

Appleton, J. H. (1975) *The Experience of Landscape*. London: John Wiley.

Barry, T. (ed.) (2000) *A History of Settlement in Ireland*. London: Routledge.

Bartlett, T. (2010) *Ireland: A History*. Cambridge: Cambridge University Press.

Bradley, R. (1998) *The Significance of Monuments*. Oxford: Routledge.

Cooney, G. (2000) *Landscapes of Neolithic Ireland*. London: Routledge.

Cooney, G. and Mandal, S. (1998) *The Irish Stone Axe Project*, vol. 1. Dublin: Wordwell.

Cunliffe, B. (2001) *Facing the Ocean: The Atlantic and Its Peoples*. Oxford: Oxford University Press.

Darcy, R. and Flynn, W. (2008) 'Ptolemy's Map of Ireland: A Modern Decoding', *Irish Geography* 41(1): 49–69.

Downham, C. (2018) *Medieval Ireland*. Cambridge: Cambridge University Press.

Estyn Evans, E. (1957) *Irish Folk Ways*. London: Routledge & Kegan Paul.

———. (1978) *Mourne Country: Landscape and Life in South Down*, 3rd ed. Dundalk: Dundalgan Press.

———. (1981) *The Personality of Ireland*, 2nd ed. Belfast: Blackstaff Press.

Flanagan, L. (2000) *Ancient Ireland: Life before the Celts*. Dublin: Gill & Macmillan.

Friel, B. (1980) *Translations*. London: Faber & Faber.

Graham, B. (ed.) (1997) *In Search of Ireland: A Cultural Geography*. London: Routledge.

Harbison, P. (1988) *Pre-Christian Ireland: From the First Settlers to the Early Celts*. London: Thames & Hudson.

Heaney, Seamus (2014) *New Selected Poems, 1988–2013*. London: Faber & Faber.

Herries Davies, Gordon L. (1983) *Sheets of Many Colours: The Mapping of Ireland's Rocks, 1750–1890*. Dublin: Royal Dublin Society.

Holland, C. H. (2003) *The Irish Landscape: A Scenery to Celebrate*. Edinburgh: Dunedin Academic Press.

Holmes, A. (1944) *Principles of Physical Geology*. London: Thomas Nelson & Sons.

Kavanagh, Patrick (2005) *Collected Poems*. London: Penguin Modern Classics.

Leeder, Mike and Lawlor, Joy (2017) *GeoBritannica: Geological Landscapes and the British Peoples*. Edinburgh, Dunedin Academic Press.

Longley, Michael (2007) *Collected Poems*. London: Jonathan Cape.

Lyle, P. (2010) *Between Rocks and Hard Places: Discovering Ireland's Northern Landscapes*. Belfast: Geological Survey of Northern Ireland.

———. (2016) *The Abyss of Time: A Study in Geological Time and Earth History*. Edinburgh: Dunedin Academic Press.

McCabe, A. Marshall (2008) *Glacial Geology and Geomorphology: The Landscapes of Ireland*. Edinburgh: Dunedin Academic Press.

Mitchell, G. F. (1990) *The Way That I Followed*. Dublin: Country House.

Mitchell, G. F. and Ryan, M. (1997) *Reading the Irish Landscape*, rev. ed. Dublin: Town and Country House.

Mitchell, W. I. (ed.) (2004) *The Geology of Northern Ireland: Our Natural Foundation*, 2nd ed. Belfast: Geological Survey of Northern Ireland.

Müller, G. (2012) *Europe's Field Boundaries*. 2 vols. Stuttgart: Neuer Kunstverlag.

Parkes, M. (ed.) (2002) *Natural and Cultural Landscapes: The Geological Foundation*. Proceedings of a Conference, 9–11 September 2002, Dublin Castle, Ireland.

Praeger, R. L. (1937) *The Way That I Went*. Dublin: Hodges, Figgis and Co.

Robinson, T. (1985) *Stones of Aran: Pilgrimage*. Dublin: Lilliput Press; repr. London: Faber & Faber, 2008.

———. (1996) *Stones of Aran: Labyrinth*. Dublin: Lilliput Press; repr. Harmondsworth: Penguin, 1997.

———. (2006) *Connemara: Listening to the Wind*. Dublin: Penguin Ireland.

Woodham-Smith, C. (1962) *The Great Hunger: Ireland 1845–1849*. London: Hamish Hamilton.

Yeats, W. B. (1992) *The Poems*, ed. A Albright. London: Everyman.

Young, M. and Donald, A. (eds) (2013) *A Guide to the Tellus Data*. Belfast: Geological Survey of Northern Ireland.

Young, M. (ed.) (2016) *Unearthed: Impacts of the Tellus Surveys of the North of Ireland*. Dublin: Royal Irish Academy.

Websites

The British Geological Survey (BGS) https://earthwise.bgs.ac.uk

Emma Cownie is a contemporary artist, focusing on the landscapes of Ireland, especially Donegal and Northern Ireland. https://emmafcownie.com

Doreen Drennan is an award-winning impressionist painter and has been documenting the Wild Atlantic Way and the Burren for over thirty years. https://doreendrennan.com

The Dry Stone Wall Association of Ireland is an organization dedicated to promoting an awareness of the craft of dry stone wall building in Ireland. www.dswai.ie/

Geological Survey of Ireland (GSI) is committed to providing free, open and accurate data and maps on Ireland's subsurface to landowners, the public, industry and all other stakeholders, within Ireland and internationally. Resources available to the public include 'Ireland Through Geological Time' (YouTube), the book, *Understanding Earth Processes Rocks and the Geology of Ireland*, and the 1:1M scale Bedrock Map of Ireland. https://gsi.ie

The Geological Survey of Northern Ireland (GSNI) provides geoscience information and services to inform decision-making. www2.bgs.ac.uk/gsni/

NASA (National Aeronautics and Space Administration) is an independent agency of the US federal government responsible for the civil space programme, aeronautics research and space research. NASA studies our own planet more than any other. They operate 26 missions in orbit and sponsor hundreds of research programmes each year. They observe oceans, land, ice and atmosphere, and measure how a change in one can drive change in others. https://science.nasa.gov/earth/

Real Map of Ireland Geological Survey of Ireland and the Marine Institute. www.marine.ie/site-area/irelands-marine-resource/real-map-ireland-0

Glossary

absolute age (or absolute time): the dating of geological events in years.

accretionary prism: an accretionary prism forms from sediments that accrete or stick onto the non-subducting tectonic plate at a convergent plate boundary.

accretionary terranes: a fault-bounded geological entity characterized by a distinctive stratigraphic or structural history, which differs markedly from the surrounding areas. A terrane is a fragment of crust broken from one tectonic plate and accreted or sutured to another plate.

adit: a horizontal or near-horizontal passage by which a mine is entered.

angular unconformity: an unconformity where horizontally parallel strata are deposited on tilted and eroded layers, producing an angular discordance with the overlying horizontal layers.

Anthropocene: the proposed most recent geological time period based on global evidence that Earth system processes are now being altered by human activity.

aquifer: a water-bearing layer of rock.

Archaean Eon: a unit of geological time spanning the period 4000 mya to 2500 mya.

artesian: groundwater under sufficient pressure to rise above the aquifer containing it.

asthenosphere: weak layer beneath the *lithosphere*, capable of solid-state flow; over which the tectonic plates slide.

Avalonia: Late-Precambrian to Lower Palaeozoic terrane that broke off from the margins of the supercontinent Godwana.

bailey: the open area inside the outer wall of a castle.

basement rocks: those below a sedimentary cover and that are igneous or metamorphic in origin.

batholith: a large mass of intrusive igneous rock formed deep within the Earth.

BCE: Before the Common Era. Used instead of BC (before Christ); with CE (Common Era used instead of AD (Anno Domini, in the year of our Lord).

bed: a distinct horizontal unit of sedimentary rock, representing a single episode of sediment deposition.

bedrock geology: the solid outcrop or subcrop that exists below soil and other unconsolidated deposits.

biostratigraphy: the part of stratigraphy that uses fossils and fossil assemblages to date rock units.

booleying: see *transhumance*.

borehole logging: the practice of making a detailed record, a well log, of the geological formations penetrated by a borehole. The practice is also known as well logging.

Cambrian System: a division of geological time in the Palaeozoic Era between 542 mya and 488 mya.

Carboniferous System: a term now including the two geological periods, Mississippian and Pennsylvanian, 359 mya to 299 mya.

cashel: *caiseal* in Irish, meaning a stone ringfort.

Cenozoic Era: a division of geological time between 65 mya and the present day.

chert: a fine-grained sedimentary rock composed of microcrystalline silica. Similar in composition to flint.

chevaux-de-frise: a defensive system of stone slabs planted in an upright position in the ground.

chronostratigraphy: that part of stratigraphy that is concerned with absolute ages.

clachan: a small village or hamlet in Ireland or Scotland.

clastic: sedimentary rock formed from particles (clasts) that were mechanically transported.

cloisters: a covered walk or open gallery along the walls of a building and forming a quadrangle. Typically associated with monastic settlements.

coal: carbonaceous accumulation of compacted vegetation.

coccoliths: minute circular plates composed of calcite, the main component of Cretaceous chalk.

constructive (divergent) plate margins: where new plate is created by divergent movement of two adjacent plates, marked by a mid-ocean ridge or continental rift.

convection cell: the circular pattern of a liquid, such as magma, rising and falling as it becomes hotter and cooler.

court cairns: court cairns are characterized as having an uncovered courtyard connected to one or more covered burial chambers.

crag and tail: a residual glacial land-form consisting of a small rocky hill (crag) from which extends a tapering ridge of unconsolidated debris (tail).

crannog: a fortified dwelling constructed in a lake or marsh in Ireland or Scotland dating from the Iron Age.

craton: the stable part of a continental interior, unaffected by contemporary orogenic activity.

curtain wall: a fortified wall around a medieval castle, typically linking towers together.

daughter atoms: produced by or resulting from the decay of a radioactive element.

decay series: the sequence of elements formed when a radioactive element such as thorium or uranium decays until the next stable atom, usually lead, is produced.

declination: the angle between geomagnetic north and geographical north at any point on the Earth's surface.

deep time: encapsulates the idea that geological time is vast and measured in millions and billions of years.

delta: a landform resulting from deposition of sediment carried by a river as the flow leaves its mouth and enters slower-moving water such as in a sea or lake.

demesne: a piece of land attached to a manor and retained by the owner for their own use:

dendrochronology: also known as tree-ring dating, is the method of dating tree growth rings to the exact year they were formed in a tree.

diagenesis: the changes that a sediment undergoes as it lithifies to become a sedimentary rock. These will include dewatering, compaction and cementation.

dimension stone: a type of stone that is cut and finished to specified sizes and shapes for a variety of uses.

diorite: coarsely crystalline plutonic igneous rock close to granodiorite in composition.

dolmen: a megalithic tomb with a large flat stone laid on upright ones, found chiefly in Britain and France. Also known as a portal tomb.

Eocene: a subdivision of the Palaeogene Period, 55.8 mya to 33.9 mya.

eon: a major division of geological time, such as the Phanerozoic, that contains the smaller units, eras.

epoch: a subdivision of geological time within a geological period.

era: a division of geological time, such as the Mesozoic, that consists of a number of geological periods.

estuarine: sediments formed in an estuary.

eustatic sea level changes: when the sea level changes due to an alteration in the volume of water in the oceans or a change in the shape of the ocean basin.

evaporite deposits: chemogenic sediments precipitated from water following the concentration of dissolved salts by evaporation. Comprise principally the minerals gypsum, halite and anhydrite.

facies: a group of sedimentary rocks and structures indicative of a given depositional environment.

Famine: The Great Famine, also known as the Great Hunger, was a period of starvation and disease in Ireland between the years 1845 and 1852. It constituted a major catastrophe and had a lasting impact on Irish society and history.

fjord: a long, narrow sea inlet, steep-sided and formed by glacial erosion.

gabbro: dark coarse grain intrusive igneous rock similar in composition to dolerite and basalt.

gamma radiation: gamma radiation or gamma rays are high-energy photons that are emitted by the radioactive decay of atomic nuclei.

granodiorite: light-coloured coarsely crystalline igneous rock compositionally between granite and diorite.

greenhouse gas: a gas that contributes to the warming of the atmosphere (the 'greenhouse effect') by absorbing infrared radiation. Carbon dioxide is an example.

greywacke: a poorly sorted sandstone containing more than 15 per cent of fine-grained matrix.

group: in stratigraphy, a lithostratigraphic unit, part of the geological record, consisting of two or more formations.

Gypsum: evaporite mineral, composition hydrated calcium sulphate.

Hadean: an informal name for the division of time between the formation of the Earth 4567 mya and 4000 mya.

half-life: in radiometric dating, the time taken for half the initial number of parent atoms to decay to daughter atoms.

hard pan: a hardened impervious layer, typically of clay, occurring below the soil and restricting drainage and plant growth. **Iron pan** is a hard pan in which iron oxides are the chief cementing agents.

Hercynian orogeny: Late-Carboniferous mountain-building episode. Also known as the Variscan orogeny.

hillfort: small hilltop settlements fortified with earthworks.

hunter–gatherers: nomadic people living chiefly by hunting and fishing and harvesting wild food.

hydrosphere: the water portion of the Earth in contrast to the solid part.

Iapetus Suture: line marking the collision of Laurentia and Gondwana as the Iapetus Ocean closed.

intrusion: a body of molten igneous rock that has intruded or invaded older rocks beneath the surface and subsequently solidified.

Iron Age: an archaeological era during which the dominant tool-making material was iron. Preceded by the Bronze Age and the Stone Age.

isostatic readjustment: the process of gravitational equilibrium at the Earth's surface in which topographic variations are balanced by density variations beneath.

isotope: one of several forms of one element, all having the same number of protons in the nucleus but with differing numbers of neutrons, and so different atomic weights.

karst: permeable limestone landscape characterized by barren rocky ground, underground drainage and an absence of surface streams and lakes.

keep: a fortified tower built within castles during the Middle Ages.

knapping: to break or chip rock or minerals with sharp blows, as in shaping flint nodules into tools.

La culture de La Tène: a European Iron Age culture, La Tène metalwork in bronze, iron and gold is stylistically characterized by inscribed and inlaid intricate spirals and interlace on bronze vessels and shields.

landscape: all the visible features of an area of land, often considered in terms of their aesthetic appeal; the appearance of that portion of land which the eye can view; the aspect of a country.

Laurentia: supercontinent comprising approximately the present continents of the Northern Hemisphere.

lazy bed cultivation: a traditional method of cultivation in Ireland and Scotland, most often used for potatoes, where parallel ridges and furrows are constructed with narrow drainage channels between them. The sod taken from the drainage trench is folded over onto the ridge thus increasing the depth of soil available and also killing any weeds in the sod.

leaching: the process of draining away of a soluble element of mineral from soil or ash by the action of percolating liquid, often rainwater.

limonite: also known as *bog iron ore*, limonite is an iron ore consisting of hydrated iron oxide and hydroxides.

lithified: a state reached after the processes that convert a newly deposited sediment into an indurated rock.

lithosphere: the strong upper layer of the Earth, including the crust and the upper part of the mantle.

lithostratigraphy: the definition of units of rocks in terms of their lithology and their correlation from area to area.

marble: the metamorphic equivalent of limestone.

member: in stratigraphy, a division of a formation, generally of distinct character or only local extent.

Mesozoic Era: the era of geological time between the Palaeozoic and the Cenozoic, occurring between 251 mya and 65 mya.

metamorphic rocks: those rocks changed by the effects of heat or pressure or both, with the development of new minerals and structures.

mica: sheet silicate minerals capable of being split into thin flexible sheets.

microlith: a small, shaped flint, typically part of a composite tool such as a spear or saw.

Miocene: a subdivision (epoch) of the Neogene period 23 mya to 5.3 mya.

motte: a mound forming the site of a castle or camp.

Neogene: a division of geological time in the Cenozoic Era between 23 mya and 2.6 mya.

Neolithic: literally, the New Stone Age, in Ireland from 3000 BCE to 1750 BCE.

Neptunist: a follower of the theory proposed by German geologist, Abraham Gottlob Werner, that rocks were formed by crystallization of minerals from the Earth's oceans.

Normans: the Norman invasion of Ireland took place during the late twelfth century when Anglo-Normans conquered and acquired large portions of land in Ireland over which the English crown then claimed sovereignty.

obelisk: a tapering stone pillar, typically with a square or rectangular cross section, set up as a monument or landmark.

ochre: a natural clay earth pigment of iron oxide and clay or sand.

Ordovician System: a division of geological time in the Palaeozoic Era between 488 mya and 444 mya.

orogeny: process in which a section of the Earth's crust is folded and deformed by lateral compression to form a mountain range.

Palaeocene: the first epoch of the Palaeogene Period in the modern Cenozoic era.

Palaeogene: a division of geological time in the Cenozoic Era between 65 mya and 23 mya.

Palaeolithic: literally, the Old Stone Age, from 12000 BCE to 8000 BCE

Palaeozoic Era: the era of geological time between 542 mya and 251 mya.

palynology: the study of fossil and living pollen, spores and other similar palynomorphs.

parent atoms: refers to the atom that undergoes radioactive decay in a nuclear reaction.

passage tomb: a passage tomb consists of one or more burial chambers covered in earth or stone and having a narrow access passage made of large stones.

peat: an accumulation of partially decayed vegetation uniquely found in bogs.

permafrost: soil, rock or sediment that is frozen for more than two consecutive years.

petroglyph: an image created as a form of rock art by removing part of the rock surface by incising, picking, carving or abrading.

Phanerozoic Eon: a major division of geological time between 542 mya and the present day. Phanerozoic means 'visible life'.

plane of unconformity: a surface of erosion or non-deposition separating younger rocks from older.

plantations: plantations in sixteenth- and seventeenth-century Ireland involved the confiscation of Irish-owned land by the English Crown and the colonization of this land by settlers from Britain.

porcellanite: rock composed of thermally metamorphosed clay-rich rocks.

porphyry: an igneous rock in which larger crystals are set in a finer-grain matrix.

Proterozoic: an Eon, a major division of geological time between 2,500 mya and 542 mya.

radiogenic elements: those elements formed as a result of radioactive decay.

radiometric dating: method of obtaining ages of geological materials by measuring the relative abundances of a radioactive element and the lighter element produced from it by radioactive decay.

raised beach: Coastal beaches and cliffs above the present shore line, in Ireland, due to higher sea levels caused by melting ice caps at the end of the Ice Age.

regression: periods when the sea level of the oceans falls and the continental shelf area may be exposed.

relative age: the dating of geological events according to the system of successive eras, periods and epochs.

ria: a funnel-shaped estuary at a river mouth formed by submergence of the lower reaches of the river valley.

rock art: human-made markings made on natural surfaces, typically vertical stone faces.

rock cycle: the succession of events that results in the transformation of Earth materials from one rock type to another.

Rodinia: a Late Proterozoic supercontinent that assembled around 1.3 bya and broke up around 750 mya.

sea-floor spreading: the theory of the generation of the ocean crust at ocean ridges and its destruction at ocean trenches.

seismology: the study of vibrational seismic waves, earthquakes and the interior structure of the Earth.

series: in stratigraphy, consists of the rocks formed during an epoch.

silica: an oxide of silicon that is found in a variety of forms, commonly as the mineral quartz, an important component of sedimentary rocks. Used by some organisms such as radiolarians and diatoms to form skeletal components.

sill: flat-lying mass of igneous rock intruded conformably with the layering of the rocks in which it has been injected.

Silurian System: a division of geological time in the Palaeozoic Era between 444 mya and 416 mya.

sinkhole: a small steep depression caused by the dissolution and collapse of underground caves in limestone areas.

slate: a fine grain metamorphic rock derived from mudstone or shale and possessing a well-developed foliation.

sphagnum moss: a genus of mosses that grow in bog areas where their remains become compacted with other plant debris to form peat.

stage: a small unit of rock contained within a short subdivision of geological time.

Stone Age: a broad prehistoric period during which stone was widely used to make implements with an edge, a point or a percussion surface. Came to a gradual end with the advent of metalworking, with the use of bronze, then iron.

strata (singular *stratum*): layers in sedimentary rocks.

stratigraphic column: a representation used in geology to describe the vertical location of rock units in a particular area. The oldest rocks are at the bottom and the youngest at the top.

stratigraphy: that branch of geology concerned with the ordering of the rock succession and its dating.

striation: scratches left on bedrock by fragments of rock embedded in overriding ice, often showing the direction of ice movement.

subduction: the process whereby an oceanic plate descends into the mantle along a subduction zone.

supercontinent: a large continental mass consisting of several components that previously, or subsequently, were continents themselves.

superficial deposits: unconsolidated residual, glacial or alluvial deposits lying on the bedrock.

system: a major division of the geological rock succession equivalent to the geological period used for the division of that time; for example, the Cambrian System covers the rocks of the Cambrian Period.

terrane: a fault-bounded area of crust characterized by a distinctive stratigraphical or structural history, implying a distinct provenance.

townland: townland (Irish: *baile fearainn*) is a small geographical division of land used in Ireland and the Western Isles of Scotland and typically covering 100 to 500 acres (40 to 200 ha).

transhumance: the traditional practice of moving livestock to upland pastures during the summer months. Known in Ireland as *booleying*.

triangulation: the process of determining the location of a point by forming triangles to the point from known points.

Triassic System: a division of geological time in the Mesozoic Era between 251 mya and 200 mya.

uniformitarianism: the theory that Earth's geological processes throughout geological history result from the action of continuous and uniform processes, acting with essentially the same intensity in the past as they do in present.

volcanic arc: a volcanic arc is part of an arc-trench complex, which is part of a subduction zone visible at the Earth's surface.

Vulcanist: a follower of the theory proposed by the Scottish scientist James Hutton that rocks were formed at high temperature deep in the Earth. This idea eventually led to the role of volcanoes in the production of igneous rocks.

Wedge tombs: wedge tombs are characterized by their sloping roof and narrowing walls at one end to produce the typical wedge shape.

Index

Page numbers in **bold** refer to figures.